Pelican Books

The Energy Question

Gerald Foley was born in Ireland in 1936. He studied engineering at University College, Cork, and at Leeds University. He worked as a professional engineer for twelve years. Since 1971 he has been engaged in research and writing about energy. From 1971 to 1979 he taught at the Architectural Association School of Architecture in London where he was Director of the Post-Graduate Energy Programme. He is presently a Fellow of the International Institute for Environment and Development (IIED) in London and a Senior Visiting Research Fellow of the Beijer Institute of the Royal Swedish Academy of Sciences in Stockholm. His publications include *Nuclear or Not: Choices for our Energy Future* (joint editor), 1978; *A Low Energy Strategy for the United Kingdom* (co-author), 1979. He was also consultant editor to the report of the World Coal Study, *Coal – Bridge to the Future*.

Charlotte Nassim studied at the Architectural Association. A mistrust of many of the assumptions made in planning and architecture led her to investigate energy availability and conservation.

Gerald Foley
with Charlotte Nassim

The Energy Question

Second Edition

Penguin Books

Penguin Books Ltd,
Harmondsworth, Middlesex, England
Penguin Books,
625 Madison Avenue, New York, New York 10022, U.S.A.
Penguin Books Australia Ltd,
Ringwood, Victoria, Australia
Penguin Books Canada Ltd, 2801 John Street, Markham, Ontario,
Canada L3R 1B4
Penguin Books (N.Z.) Ltd,
182–190 Wairau Road, Auckland 10, New Zealand

First published 1976
Reprinted 1976
Reprinted with an index 1978
Second edition 1981

Set, printed and bound in Great Britain by
Cox & Wyman Ltd, Reading
Set in Monotype Times

To Lanna, Katie, and Conor

The laws expressing the relations between energy and matter are not solely of importance in pure science. They necessarily come first . . . in the whole record of human experience, and they control, in the last resort, the rise or fall of political systems, the freedom or bondage of nations, the movements of commerce and industry, the origin of wealth and poverty and the general physical welfare of the race. If this has been too imperfectly recognized in the past, there is no excuse, now that these physical laws have become incorporated into everyday habits of thought, for neglecting to consider them first in questions relating to the future. – FREDERICK SODDY, *Matter and Energy*, 1912

Contents

Acknowledgements

I am grateful for the help and advice given to me by many friends and colleagues during the preparation of both editions of this book. Among them I would especially mention Colm Foley, Adrian Cottrill, John Thomas, Peter Chapman, Leonard Brookes, Philip Green, George Kasabov and John Williams.

Gerald Leach at IIED, who suggested the book in the first place, has helped throughout with advice, information and encouragement. Ariane van Buren, also at IIED, has been an indispensable colleague on a daunting series of projects over the past four years. To both of these my very special gratitude is due.

John Seed kindly allowed me to use Figure 6 which is based on his work at the Architectural Association. I would also like to pay an affectionate tribute to my students at the Architectural Association. Their work has considerably enriched the content of the book and the discussions around which it was built.

I would also like to thank Måns Lönnroth at the Secretariat for Future Studies in Stockholm. I have repaid his generosity with his ideas by shamelessly adopting many of them as my own. I am also grateful to Gordon Goodman and Lars Kristoferson at the Beijer Institute in Stockholm; to Carroll Wilson and my many friends from the World Coal Study; and to Sir George Porter for his help and encouragement during my work on two Energy Forums at the Royal Institution.

Typing from the final manuscript of the first edition was a major exercise carried out with cheerful efficiency by Carol Hogan, Jo Olsen and Mrs B. M. Alcock. Alison Forbes was immensely helpful in getting the final manuscript into shape.

My wife and family have stoically endured all the miserable repercussions of having a book written and rewritten in the house. When it all became too much my mother and grandmother provided a retreat in the peace and calm of Sligo.

No one, except myself, should be blamed for any deficiencies of style or content in the final work.

G.F.

Notes on Units and Conversions

No widely used system of units covers the full range of topics discussed in this book and any attempt to impose one would be both pedantic and confusing. The following conventions have, rather arbitrarily, been chosen for use in the text; tables of detailed conversion factors have been included in the Appendix.

Metric System

This has been used throughout with the exception of oil-industry statistics which obstinately follow rules of their own. The small difference between Imperial tons and metric tonnes has generally been ignored.

Numbers

Powers-of-ten notation has been used for very large and very small numbers. A million is thus $1{\cdot}0 \times 10^{6}$ and a millionth is $1{\cdot}0 \times 10^{-6}$. American usage has been followed for the term 'billion' which is taken as a thousand million or $1{\cdot}0 \times 10^{9}$.

Oil-Industry Statistics

The oil industry almost invariably expresses quantities of petroleum products in barrels. The barrel is a measure of volume and the number of barrels per tonne varies in accordance with the density of the product – from 6·7 barrels per tonne in the case of heavy crude oil, up to 8·5 barrels per tonne in the case of motor spirit. The commonly accepted average figure of 7·3 barrels per tonne has been used for crude oil and petroleum products in general.

A barrel of oil contains almost 160 litres, 35 Imperial gallons or 42 American gallons.

Energy Equivalents and Statistical Inconsistencies

Oil, gas and electricity are often expressed in tonnes of coal or tonnes of oil equivalent. The conventions by which these conversions are made vary widely and caution is required when comparing statistics from different sources. The Appendix gives details of some of the conventions used by different data-producing agencies. No attempt has been made to 'correct' or harmonize the various sources used in the book. To do so would introduce even further distortions and inconsistencies into statistics which, at best, are only likely to be accurate to within 5–10 per cent.

Some 'Energy Signposts'

The kilowatt-hour (kWh) is the energy unit most commonly used in the book. The following approximate figures may help readers obtain a sense of the comparative magnitude of various energy-using processes:

A single-bar electric fire uses a kilowatt-hour every hour.

The daily intake of food energy of a well-nourished western adult male is about 3 kilowatt-hours; the maximum daily output of work of a manual worker is about 0·5 kilowatt-hours. The work output of a horse in an eight-hour day is about 6 kilowatt-hours.

A litre of petrol contains about 10 kilowatt-hours.

A tonne of coal contains about 8 000 kilowatt-hours.

A typical large modern coal-fired power station (of 1 000 megawatts) produces a million kilowatt-hours every hour and consumes about 400 tonnes of coal while doing so.

The United Kingdom consumes about 340 million tonnes of coal equivalent every year.

The world uses about 10 billion tonnes of coal equivalent every year.

The total amount of solar energy intercepted by the earth every hour is about $1{\cdot}7 \times 10^{14}$ kilowatt-hours, more than twice the total annual energy consumption of the world.

Preface to the Second Edition

The first edition of this book was published nearly five years ago. A great deal has happened since then to maintain energy as a topic of intense concern. Anti-nuclear movements, queues at filling stations, doubling fuel bills, and the almost daily headlines about events in oil-producing countries combine to keep us aware of its importance. Governments now have ministries of energy, newspapers have energy correspondents, and libraries have shelves of energy books.

Surprisingly little, however, has changed in the way the world uses energy. Patterns of energy consumption are almost exactly the same as they were five years ago; the total consumed continues to creep slowly upwards. The fundamental change is to be found in the new recognition of energy problems; they are no longer thought to be temporary or trivial.

At the time when the first edition was being prepared data on energy use and descriptions of energy technologies were not easy for the non-technical reader to obtain. Now there is no such problem – rather the contrary. The difficulty is finding a path through it all.

The first edition endeavoured to act as a source of information and a guide to what was significant. Some of its arguments have proved themselves gratifyingly durable. The view that world oil consumption had more or less reached its peak, somewhat heretical at the time, has now been generally accepted. The inevitability of a relatively minor role for nuclear power, as a global energy source, for at least the remainder of this century is also now accepted even by the nuclear industry itself. The pessimism about alternative energy sources is, if anything, reinforced by more recent information.

Other judgements were less penetrating. The true potential and importance of energy conservation was not recognized. Conservation is, in fact, the key to an energy strategy capable of sustaining industrial society or providing a possibility for an orderly transition to some other kind of society.

A few of the interests of the time, energy analysis for example, were given a prominence they did not deserve. The discussion of the relationship between Gross Domestic Product and energy consumption failed to see the obvious.

The objectives of this edition remain those of the first: to provide information and a guide to what is most important. The balance of topics has been somewhat altered and the emphasis on the United Kingdom has been reduced. The data in the tables have been extended and brought up to date.

The biggest omission is an adequate analysis of energy in the developing countries. Their energy plight is worsening and is inextricably linked with their problems of urban and rural poverty, population growth, and finding appropriate and sustainable development strategies. They are topics which take us far beyond the scope of this book. Suffice it to say that the industrialized countries use four-fifths of the world's energy and dictate the pattern of consumption of the rest. If the developed countries cannot solve the energy problems they are going to face in the coming decades, there is little hope for a peaceful future anywhere in the world.

G.F.
C.N.
1980

Introduction

Energy is one of our most familiar concepts. People talk of the energy of a lively child, of the sun, of the waves, or of the wind. They recognize that energy is stored in a spinning flywheel, in the wound spring of a clock or in the plutonium core of a nuclear bomb.

William Blake said, 'Energy is Eternal Delight'. The cautious author of a book on thermodynamics ventured: 'Energy may be thought of as a capacity for doing work.' And Einstein pointed out that energy is equal to mass multiplied by the velocity of light squared.

Energy is all of these and more. It is described in a variety of ways and none defines it fully. Each reveals an aspect of energy, the one appropriate to the viewpoint of that particular time. Energy is heat, light, electricity and a capacity to do work. More fundamentally, it is that which each of these has in common with the others. And energy is matter. It is this last which enables it to be said that energy is everything, the ultimate, irreducible essence of the universe.

But considering universal truths is useful only up to a point. Looking at everything can prevent one seeing anything, particularly the practical. And energy is emphatically a practical concern. Energy grows food and keeps people alive. It transports them, fuels machines and sustains economic systems. The fossil fuels – coal, oil and natural gas – are the basis of modern industrial society. When energy supplies are in question all that depends on them is also in question; not only the way of life but life itself.

A chilly air of reality obtrudes itself very quickly into any discussion about energy. Food and warmth, transport and jobs, standards of living and future prospects, all depend on society's ability to continue supplying itself with the energy it has grown to need. It must do this without disrupting the great natural energy systems of the earth on which the heat balance and climate depend, and without destroying the living web of the biosphere of which man is himself a part. And it must look beyond these immediate concerns to the inevitable decline in the availability of the fossil fuels and the need to find a substitute for them if industrial civilization is to survive.

Modern society has grown without understanding the full nature of its dependence on energy. The great, complex edifice of industrial civilization, with its apparently endless catalogue of achievement and conquest of the physical world, has so impressed or bewildered people that they have failed to see the fragility of its supports. Victorian Britain was not just a pyrotechnical display of mechanical and scientific inventiveness, commercial skill and opportunism, imperial ambition and military prowess. It was also a twenty-fold increase in the consumption of coal. America's ability to put a man on the moon sprang not just from the skill of its scientists and technologists; it was also a product of a society which in 1970 was consuming a total of thirteen times as much energy as Britain was at the height of its imperial power just before the First World War.

Man's discovery of how to make the deserts bloom, to win two harvests where previously there had been one, to break and manipulate to his own advantage the genetic codes of plants and animals, are all part of the remarkable achievements of modern agriculture. They are a reason, to many, for self-congratulation and optimism about the earth's ability to support an apparently never-ending increase in population. Purpose-bred cows now produce a hundred times as much milk as their forebears; battery-reared chickens function like docile egg-machines; and the farmer, with a tiny proportion of the labour, often produces many times the crop yields of a few generations ago.

Man, however, has not outsmarted evolution and made nature

inherently more efficient. The increased yields are possible only because he has learned how to provide an energy subsidy to the natural processes of food production. This is not to say that an energy subsidy is a bad thing. Man cannot eat coal or oil. If, by his ingeniousness, he can channel them into the agricultural system and increase its yields he is effectively creating an additional food source for himself. But it is important that he should not delude himself into believing that he has made permanent improvements in nature. Without his care and energy subsidies, his specialized food producers – chickens, cows, 'green revolution' rice strains and the rest – are incapable of survival. The steadily diminishing fossil-fuel resources are now an essential part of humanity's food supply.

The size and shape of its cities clearly demonstrates the amount of energy a society has at its disposal. There is a modern, often sentimental, attachment to what is old and quaint: the tight narrow streets of a medieval town, or the compact cluster of shops, houses, pub, post-office and library round a Victorian railway station. These have an intimacy and human scale lacking in the multi-storey office blocks and acres of car-parking typical of modern urban development. But it should be recognized why the older city was small and compact. It is unlikely previous generations were less megalomanic than the present; they may even have been more so. Cities were built the way they were, not because of any attachment to the small and intricate for its own sake, but because it was the only way to make them work with the amounts of energy available at the time.

For most of human history the majority of people have had nothing but the energy of their own leg muscles to transport themselves. In the medieval city, therefore, all the necessities of daily life had to be located within walking distance of everyone. Shops, markets, workplaces and dwellings were clustered together with a minimum wastage of space. Furthermore, the slow pace of animal carts and beasts driven along unmetalled roads limited the area of hinterland from which a city might draw its food supplies. The population of the city was restricted to that which could be fed from the food produced within, at most, a couple of days' journey by animal cart.

But man learned to harness new sources of energy and use them for transport. And as transport systems developed the size of cities could increase. Ship-borne grain and other food could supplement that produced locally. Well-made roads and improved vehicles could extend the usable hinterland. Yet as late as 1830, the London villages, now more familiar as the names of stations on the Underground railway, were being supplied with food carried in on carts or driven on foot from the fields around them. The size of a town and its supportable population was, even then, governed by such factors as how far one could make a cow or a goose walk, and the length of time food would remain edible as it was being transported.

Now, the abundant energy of petroleum has removed most of these limitations. Food can be carried hundreds of miles by road or rail in a single day. Ships make the rich food-producing areas of the world accessible to everyone. Refrigerated storage eliminates the problem of decay. A country with money to spend can obtain all the food it needs to feed its people at any level of nutrition they care to adopt. A measure of the change which has occurred can be obtained by a simple comparison: a tonne of oil contains an amount of energy equivalent to the hourly output of a herd of no less than 16 000 horses. A couple of modern trucks are more effective than the whole transport system of even the largest cities of 150 years ago.

Buildings, too, are a commentary on the availability of energy. The tall glass-walled office-block or hotel defies nature. Left to itself it would be unusable. It absorbs heat like a greenhouse when the sun shines and loses it as quickly when the weather turns cold. But energy is pumped into it to keep it cool in summer and heat it in winter. It is designed to be independent of what is happening outside, a self-contained capsule isolated from the vagaries of the external climate. These buildings can look identical wherever they are erected because their heating and cooling equipment make them capable of transcending the variations in climate between one place and another. But they use a great deal of energy in doing so. The annual consumption of a modern air-conditioned office block in the UK can be about ten times as high, per unit area, as that of a comfortable house.

The influence of energy availability is seen in the form and structure of society itself. If the majority of its people must spend most of their time working to obtain the food necessary for their own survival, a society will have the characteristics of a subsistence agricultural economy. Its superstructure of government and culture will be simple. Its main concerns will be the development of ways of dealing with the periodic assault of natural disasters, keeping the population within sustainable limits, and devising a system of myths and rituals which will allow it to come to terms with the mysteries of the world about it.

The gulf between the subsistence economy and the highly developed one does not exist because the researcher examining samples of moon-rock is inherently superior to the hill shepherd. It depends almost entirely on the relative availability of usable energy within the two societies. The average American directly and indirectly consumes over 300 times as much energy as the average Ethiopian. Like some great hovercraft, industrial society is lifted and maintained above concern with the elemental necessities of life by a prodigious expenditure of energy. Without this energy, the skills of industrial man are merely a burden in the struggle for survival.

The theme of this book is the relationship between energy use and the evolution, present functioning and potential future of human society. Its aim is to develop an 'energetic' viewpoint, a way of seeing human activities as energy-using processes depending entirely on the continued availability of energy resources.

The first part examines energy systems. Human life depends on a delicate balance between the earth's absorption and its reflection of solar energy. The distribution of deserts and fertile areas, ice-bound tundra and temperate zones, even the relative areas of land and ocean, depend on the energy-driven circulatory systems of the winds, rains and ocean currents.

The history of man's emergence from the ecological niche appropriate to a medium-size omnivorous mammal to his present position as the earth's dominant species is one of increasing skill in harnessing and manipulating energy. Each stage of his evolution has been marked by an extension of his ability to control the natural flows and accumulated energy resources about him.

But energy has been more than just the fuel for man's material advance; it has been a major unifying principle in the development of his understanding of the universe. From the beginnings of modern scientific thought in the Renaissance, through to modern physics, the concept of energy has been steadily widened. The fundamental principles of heat, light, sound, electricity, magnetism and wireless were brought together in the magnificent work of discovery and synthesis of the eighteenth and nineteenth centuries. Energy was the common ground on which these independent scientific disciplines could be joined and mutually reinforced. The twentieth century brought Einstein and a revolutionary proposition about the interchangeability of matter and energy. No more comprehensive satisfactory hypothesis linking the whole range of scientifically measurable physical phenomena has yet emerged.

In the early 1900s Frederick Soddy speculated on the nature of the dependence of human social systems on energy consumption and suggested that the functioning of society was governed by the same natural laws as all other energy systems. He foresaw a time when the laws of energy would provide an intellectual foundation for economics and sociology. In recent years the American ecologist, Howard Odum, has examined the energy flows of natural eco-systems, refined by the trial and error of billions of years of evolution, in the belief that the principles by which they operate will be found applicable to the energy-using system of human society. Such dreams of a mechanistic system of values by which human social behaviour might be governed are, fortunately, unlikely to be fulfilled. But there is much to be learned still about how the functioning of society is affected by the type and magnitude of the energy flows which sustain it.

The second part of the book looks at the extent of the energy resources on which man depends. Coal, oil and natural gas are the basis of industrial society but they are disappearing at an extremely rapid pace. The question of how long these resources can last and how they will be shared is obviously important.

Many people are optimistic about possible alternatives to these fuels. The world has vast untapped hydro resources. There are huge quantities of oil trapped in tar sands and oil shales. Enthusiasts talk of obtaining energy from the sun, the wind, the waves, the earth's internal heat, the temperature gradients in the deep oceans, or of turning plant material into liquid or gaseous hydrocarbon fuels. There are the promised benefits, and recognized dangers, of nuclear power. There is the glittering dream of limitless energy if the fusion reaction of the hydrogen bomb can be harnessed for peaceful use.

But in looking at the earth's energy resources it is necessary to go beyond the mere question of their absolute magnitude, or their theoretical potential if the technology to harness them could be devised. The practical availability of energy is limited by social, geographical, political, economic and technical constraints. The fact that the Middle East contains 60 per cent of the world's proved oil reserves has considerable implications for the future pattern of oil consumption. Proving the existence of a resource does not make it accessible to everyone. Recognizing the potential of nuclear power does not make it possible to construct trouble-free nuclear power stations. Demonstrating the need for an alternative source of supply does not necessarily mean that the technical and environmental problems of obtaining oil from coal or shale can be quickly or satisfactorily solved.

The third part of the book is called 'Futures'. It does not attempt to make detailed forecasts of what is going to happen. Rather it discusses how society may begin to recognize the limits and constraints imposed upon it by energy availability. Knowledge of these is necessary before it can make rational choices in the process of creating its own future.

What is called 'planning' is too often wishful thinking. It refuses to recognize limits. Past trends are projected into the future as demands which, somehow, someone, somewhere, will produce the resources to satisfy.

But there are limits and the universe owes no society a living. Planning the best future is a matter of seeing these limits and making the best use of the resources available within them. The limitless dreams of just a few years ago will never be fulfilled; that much, at least, is now clear. It does not mean that nothing can be done but that more thought is required to establish what can be done. Wasteful uses can be eliminated to leave resources for productive purposes – in other words, energy can be conserved. Conservation has a bad name with connotations of deprivation and enforced frugality. It needs to be seen differently, as creating opportunities.

If new opportunities are to be created within a restricted future, realism is a necessity. So also is imagination. The opportunity to use less energy more wisely and thereby achieve advances in material living standards, particularly for those at present poor, cold and hungry, still exists.

It would, however, be both naïve and presumptuous to minimize the difficulties likely to be encountered. Human society is complex, both vulnerable and resilient in surprising ways, and as totally dependent on energy as it is on food. As energy supplies become curtailed there will be problems and a need to make difficult choices. The work of preparing for this is urgent and important. It has no place for the arrogance of facile optimism, or pessimism.

Part One

Systems

1

The Earth's Energy Systems

In the sun-centred myths and rituals of many of his early religions man acknowledged the supremacy of the sun-god who controlled night, day, the seasons and elements. Later cosmologies, founded on an anthropocentric concept of the universe, relegated such ideas to the realm of childish superstitions and asserted the primacy of man. But modern science has firmly reinstated the sun. All living creatures depend on it. It provides the energy without which nothing could grow and nothing could feed. As ancient man believed, it is, indeed, the source of the energy of life.

The energy of the sun originates in its core where the temperature is about 12 000 000°C. Under these conditions matter does not exist in familiar forms. Atoms are stripped of their electrons and elements are transmuted as the nuclei of light atoms fuse to form heavier elements. In the process some mass is transformed into energy in accordance with Einstein's equation $E = mc^2$ (energy is equal to mass multiplied by the velocity of light squared). The sun is losing mass at the astonishing rate of six million tonnes a second.

The energy released by this nuclear fusion is in the form of electromagnetic radiation. This is a generic description of all radiant energy. It includes visible light, X-rays, wireless and television waves, ultra-violet and infra-red rays, each of which differs in wavelength from the others. Wavelengths vary from the shortest, measured in millionths of a centimetre and called gamma-rays, to the hundreds of metres of long-wave radio.

In the core of the sun, energy exists mainly in the form of gamma-rays. But at the surface, 500 000km away, the temperature is about

5 800°C and by the time the energy has reached it the radiation is no longer concentrated in the extreme short end of the electromagnetic spectrum, but has been spread and shifted along it. At the surface of the sun a large proportion of the energy is emitted in the form of light – those wavelengths visible to the human eye.

Further from the sun the concentration of energy falls. At the outer planets it is so low that it is not sufficient to support life in any of its terrestrial forms. Beyond the solar system the sun's energy is finally dissipated in the wastes of intergalactic space. But at the edge of the earth's atmosphere the intensity of radiation is still high. It amounts to 1·4 kilowatts per square metre and is referred to as the solar flux or the solar constant. The earth's outline intercepts energy at the rate of about $1{\cdot}7 \times 10^{14}$ kilowatts which is about 100 000 times the total generating capacity of all the world's electrical power stations.

Yet in spite of this inundation of energy the earth's temperature remains remarkably constant. The inflow of energy must therefore be balanced by an equal outflow. If the earth were not radiating back into space the same quantity of energy as it receives, it would be growing hotter or colder. At times in the earth's history the equilibrium between inflow and outflow has been upset. Changes in the composition of the atmosphere or in the reflective qualities of the land and water surfaces have altered the proportions of energy reflected and absorbed. The temperature has risen and the climate has become hot and arid; or it has fallen, bringing about an ice age.

When these climatic changes take place a new equilibrium is eventually reached and the energy inflow and outflow again balance. This is because the amount of energy a body radiates is related to its temperature: the higher the temperature the greater the rate at which energy is lost. If the earth begins to absorb more and reflect less energy its temperature rises. The rate of energy outflow increases until it balances the inflow. But the equilibrium temperature is now higher than it was before and life which evolved under the former conditions may find the new equilibrium intolerable. The interests of mankind would best be served by the continuation of the present division between the amount of solar energy absorbed and that reflected.

About 30 per cent of the solar flux is reflected, but the amount varies greatly between different parts of the earth. Clouds and atmospheric particles reflect some and more is reflected back from land and water surfaces. Much depends on the ground cover in a particular region. An ice-covered continent, a jungle and an ocean all reflect differently. Alterations in land use, changes in the vegetational cover, or large-scale discoloration of the seas will therefore change the present balance between absorbed and reflected energy.

No one yet knows how important man's interference with the division between reflected and absorbed solar energy has been. Certainly it is possible to change the climate: it has happened many times in the past for reasons unconnected with man's activities. Over the past couple of hundred years, and most particularly in recent decades, man has changed the texture and colour of the landscape; he has poured gases and dusts into the atmosphere; he has created clouds of smoke and steam; he has changed the colour of large areas of the seas and oceans. Some of his activities counter-act others; some reinforce each other. The results of this global experiment are completely unpredictable in the present state of knowledge. International collaboration is needed to ensure that the effects of activities, which might be harmless if restricted to a few locations, do not accumulate to a dangerous extent globally through each country independently following its own internal interests.

The remaining 70 per cent of the total flux, which is not reflected, is absorbed into the energy processes of the earth. Some of it is absorbed as it passes through the atmosphere. About 25 kilometres above the earth there is an atmospheric layer in which ozone occurs. Ozone is a molecular form of oxygen with three atoms instead of the usual two. It is important here because it absorbs ultra-violet radiation which is dangerous to life even in moderate doses; in laboratories it is often used as a sterilizing agent. If the full intensity of the ultra-violet component in the sun's radiation were to penetrate to the earth's surface most life would be eliminated.

In the ozone layer a complicated process occurs in which 'ordinary' oxygen and ozone interact with ultra-violet radiation and prevent most of it reaching the surface. There is now scientific evidence which suggests that supersonic aircraft operating at heights which bring them into the ozone layer can impair its action as a life-preserving filter. There is also a growing concern about the fluorocarbon gases released from aerosol containers: these gases are now believed to ascend into the ozone layer and damage it, though, as always, there are those who feel there is no cause for alarm. It is an issue on which the human race cannot afford to be optimistic – and wrong.

The small quantity of ultra-violet energy which does penetrate to the earth's surface can, however, be thanked for holiday suntans. When exposed to small doses of ultra-violet radiation the skin produces a pigment to protect the delicate inner tissues of the body. Even in the comparatively slight exposures necessary to produce a skin tan, ultra-violet radiation is to be treated warily: it causes some forms of skin cancer.

As it passes through the atmosphere solar radiation is absorbed by water droplets, dust particles and carbon dioxide. Some wavelengths are more heavily absorbed than others. Figure 1 shows a typical distribution of solar energy at the earth's surface in comparison with that at the outer edge of the atmosphere: note that the energy in some wavelengths of infra-red radiation has been completely absorbed. The diagram shows clear-sky conditions. Downwind from a cement works on a cloudy day in November, the solar-energy distribution at ground level would show a near zero intensity for all wavelengths.

The total radiation reaching the earth's surface is called the 'insolation'. About half of it is in the visible band of the spectrum on a clear day. All the radiation, visible and invisible, which is not reflected by the objects on which it falls heats them. During the early

Figure 1. Distribution of solar energy intensity along the electromagnetic spectrum

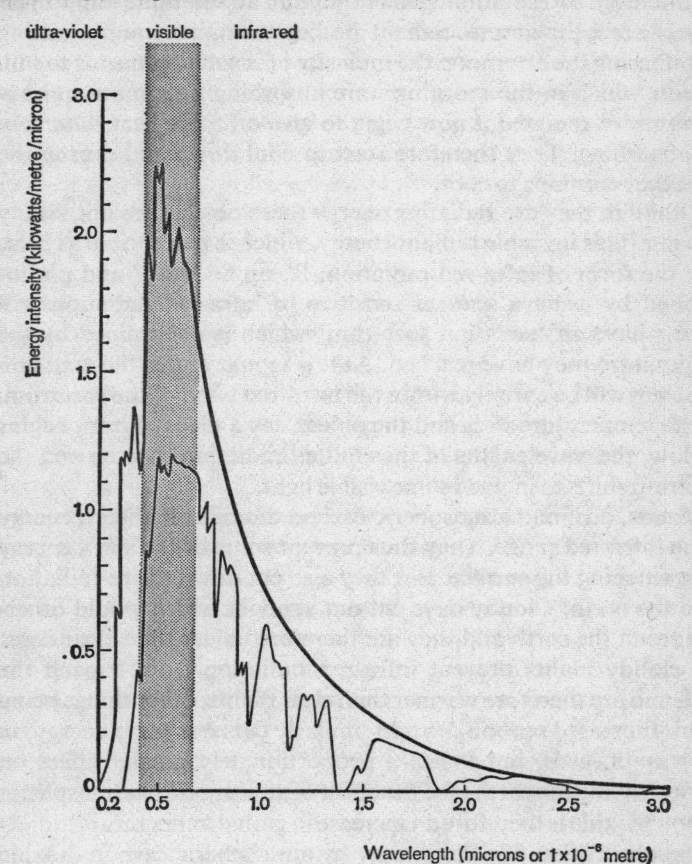

Note: Top curve shows intensities at the edge of the atmosphere. Lower curve shows typical 'clear-sky' distribution at ground level. Visible spectrum is shown shaded.

part of the day objects become warmer as they absorb energy and the intensity of radiation builds up. But at the same time these objects are behaving as radiant bodies themselves and emitting heat. During the afternoon the intensity of insolation begins to fall. Objects which in the morning were absorbing heat more quickly than they re-radiated it now begin to give off more heat than they are absorbing. They therefore start to cool down and during the night they continue to cool.

Although they are radiating energy these objects are not visibly glowing. This invisible radiant energy, which is perceptible as heat, is in the form of infra-red radiation. It can be 'seen' and photographed by using a scanner sensitive to infra-red radiation. All objects have an 'emission spectrum' which is determined by the temperature they have reached. At low temperatures the emission spectrum will be entirely within the infra-red band of the spectrum. As the temperature rises and the object, say a piece of iron, begins to glow, the wavelengths of the emitted radiation shorten and the spectrum shifts to include some visible light.

Clouds, dust and atmospheric carbon dioxide all absorb energy in the infra-red range. They thus prevent some of the sun's energy from reaching the surface. But they also cut down the re-radiation from the earth. Cloudy days cut out sunlight which would otherwise reach the earth and they are therefore colder than clear days. But cloudy nights prevent infra-red radiation from leaving the earth and are therefore warmer than clear nights, other things being equal. Increased carbon dioxide similarly cuts out some energy on the way inwards, but it has a proportionately greater effect on energy leaving the earth: the net effect of an increase in atmospheric carbon dioxide is therefore an increase in global temperatures.

The absorption of solar energy by atmospheric carbon dioxide has attracted attention in recent years because it is affected directly by human activities. Carbon dioxide is emitted in the breathing or 'respiration' of living creatures, both plants and animals. It is also emitted in the burning of wood, coal and oil. It has been estimated that the effect of the present rate of combustion of organic material is to raise carbon dioxide content of the atmosphere by about 0·4 per cent per year. In the past century the average atmospheric content of carbon dioxide has increased by about 15 per cent, but whatever difference this has made is not yet definitely identifiable. Some scientists have calculated that a doubling of the carbon dioxide content of the atmosphere could cause average temperatures to rise by about 2°C. This could be sufficient to produce considerable climatic changes.

During the carboniferous geological period carbon dioxide levels are thought to have been about twice as high as today's. This resulted in higher temperatures and better conditions for plant growth. The great carboniferous forests flourished and much of the earth's coal was laid down. It would be ironic if the combustion of that coal should once more create the ideal conditions for the formation of coal.*

Superimposed on the diurnal cycle of absorption and re-radiation of solar energy there is another, slower, seasonal rhythm. Summer heat is carried over into autumn and winter. In temperate climates the coldest weather does not come until the heat built up during the summer has been exhausted from the land and seas. November is rarely as cold as February, though they are equidistant from the time of minimum solar radiation at the winter solstice.

This pattern is modified by geographical location. The sea loses heat more slowly than the land. Hence areas close to the sea have climates which are less extreme than those of similar latitude but further inland. Towards the equator the insolation does not vary greatly between winter and summer and the seasons are almost uniform, whilst the poles have sunless days at midwinter. In each area the vegetation and the creatures which feed on it are characteristic of the pattern of solar-energy availability.

Not all of the energy absorbed by the earth, however, is radiated directly back into space. About a third of it takes one or other of a number of indirect routes on its inevitable way to the heat sink of

*See pp. 105–6.

outer space. It drives the evaporation and precipitation cycle; the winds, waves, and ocean currents; and the photosynthetic processes of the biosphere.

By far the largest of these systems is the evaporation and precipitation, or hydrological, cycle. About 23 per cent of the total solar energy intercepted by the earth goes into driving this vast engine. Most of it is used to turn water into water vapour. Once this has happened the winds transport the water vapour until it finally condenses and returns to the earth as rain or snow.

The working of this energy cycle is complex and variable in its timing. Water may condense and fall back immediately into the ocean from which it has evaporated. It may travel in a regular seasonal pattern and bring a rainy season predictable within a few days; it may produce the vagaries of an English summer. It may fall in the uplands of a continent and travel thousands of miles before it reaches the ocean again. It may become locked in a glacier or a polar icecap and remain there for hundreds of thousands of years. It may become trapped as ground-water, or interact with minerals in the earth's crust and be released only on a timescale measured by the emergence and disappearance of land masses. But in each case the energy process is the same: when water is lifted from the surface by evaporation, energy is absorbed and when it returns to the ocean all that energy has been released and the energy account has been balanced.

The interacting winds, waves and ocean currents are a much smaller energy system. They use only about 0·2 per cent of the total solar energy reaching the earth, less than a hundredth that of the hydrological cycle. The variation in the intensity of insolation over the surface of the earth causes the winds to develop. Warm air rises at the equator and draws in cooler air from areas to the north and south. The air from the equator, having lost its heat, descends at about latitude 30°, setting in motion a subsidiary system between 30° and 60°, and a further system between 60° and the polar regions. When the effects of the earth's rotation are added to these vertical circulatory movements the patterns of the basic wind systems are established. But the detailed behaviour of the atmosphere, its high-level jet-stream winds with speeds of over 500 kilometres per hour and the great eddies of the cyclones and anti-cyclones, is far from being understood. Only where there are long reaches of open ocean and freedom from the distorting effects of land-masses do the wind patterns begin to conform to those of the simple model.

The wind systems in their turn drive the ocean waves and the ocean currents. The Gulf Stream is probably the best known; others are the Humboldt current which brings cold water up along the west coast of South America, and the complex equatorial currents of the Atlantic and Pacific. The movements of the winds, waves and ocean currents are all gradually slowed down by friction and their energy is dissipated in a slight warming of the atmosphere and the oceans.

While the winds and ocean currents are tiny in comparison with the hydrological system, they have a very important role in the earth's climate. Europe would be much less habitable without the ameliorating effect of the warm waters of the Gulf Stream. The winds, by bringing rain to one area and not to another, determine the difference between desert and fertile land. The greater energy of the hydrological cycle is required to evaporate water and raise it from the oceans, but the winds determine where it falls. The major energy system is, in effect, controlled by the minor system.

All this should make self-evident the reasons why meteorologists are alarmed by proposals to dam the Bering Straits or alter the great river systems of the USSR: they would introduce a new element into the intricately meshed energy systems of the earth. To predict the results is impossible, but the range of possible effects is clear. It could alter the timing of rainy seasons; it could reduce or increase the polar icecaps, affecting the world's climate, flooding or stranding the world's coastal cities; it could create deserts on now fertile land.

The energy of the tidal systems is surprisingly small. It is only about 1 per cent that of the wind, waves and ocean currents. The tides are driven by the gravitational forces and planetary motions of the sun and moon in relation to the earth, with the moon's effect being dominant. Tidal energy is obtained at the expense of a slight slowing down of the earth's rotation – about 0·001 seconds per rotation per century. Tidal energy, too, is dissipated in friction.

The internal heat of the earth (or 'geothermal' energy), flowing outwards from the molten core, sometimes showing itself as hot springs and volcanoes, is another minor energy system. The total quantity of this energy is $3{\cdot}2 \times 10^{10}$ kilowatts, that is about ten times that in the tides, but it is only 0·2 per cent of the total solar flux. Deprived of the sun, the earth would be a cold, still, barren place.

Moreover it would not support the most intricate and delicate energy system of all: that of the living creatures of the biosphere. The energy-capturing process on which they all depend is photosynthesis, which literally means putting light together. Green plants on land and algae in water are able to use light from the blue and red bands of the visible spectrum to break down carbon dioxide and water and recombine the constituents, forming the complex carbohydrate molecules of living tissue. Their energy budget is comparatively tiny, around $4{\cdot}0 \times 10^{10}$ kilowatts, about a tenth that of the winds and ocean currents. (See Figure 2.)

The term 'biosphere' is used to define the zone encircling the earth in which life exists. Its outer limit is the upper atmosphere where occasionally bacteria and spores of fungi are blown by the wind. Its inner limit is the dark sparsely-inhabited water of the ocean depths. Only a tiny proportion of the earth's life inhabits these extremes. Most of the living creatures of the biosphere are found within a few metres of the surface of the land and in the illuminated upper regions of the seas and lakes.

The photosynthesizing organisms are the basis of life for all other living creatures. They are the only forms of life which can exist without eating other creatures and they are sometimes called 'autotrophic' or self-nourishing. The rest are called 'heterotrophic': they are nourished by others. Heterotrophic creatures include microbes, bacteria, grazing insects and animals, fish and flesh-eating animals, and man. Fungi, too, are heterotrophic

Figure 2. Distribution of solar flux between the earth's energy systems

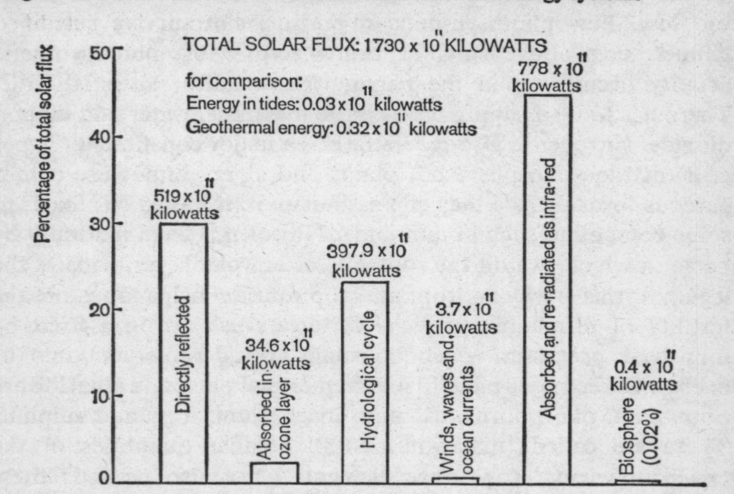

Source: M. King Hubbert, 'The Energy Resources of the Earth', *Scientific American*, September 1971.

because they can grow without light, taking their energy from decaying vegetable matter – thus making a mushroom more like an animal than a plant.

In the long-running debate about the origins of life on earth, most evidence suggests that it started in the 'primeval soup' of the oceans about three billion years ago. It had protection there from the ultra-violet wavelengths in the sun's radiation. In developing, life also created the conditions for its own wider existence. Oxygen is one of the by-products of photosynthesis. As life spread in the oceans oxygen accumulated in the atmosphere, the ozone layer began to play its protective role and the colonization of the land became possible. Land-based life probably began between 300 and 400 million years ago.

Photosynthesis needs a quite precise combination of conditions before it can occur. The temperature must be neither too high nor too low. Few photosynthetic organisms can survive sustained temperatures above 40°C or below zero. Most photosynthetic activity occurs within the narrow temperature range 10–35°C. There has to be a supply of the basic materials, water and carbon dioxide. Nitrogen is also required. It is a major constituent (78 per cent) of the atmosphere but plants and algae cannot use it in a gaseous form. Before they can assimilate it, it has to be 'fixed' in some compound such as ammonia. Nitrogen is fixed naturally by bacteria which live on the roots of certain plants, principally the legumes; this is why appropriate crop rotation helps to restore the fertility of farmland. Artificial fertilizers use nitrogen fixed by industrial processes which consume considerable amounts of energy. Other major mineral requirements of plant and algal life are potassium, phosphorus, calcium, magnesium, iron and sulphur. These are called 'macro-nutrients'. Smaller quantities of the 'micro-nutrients' (or 'trace elements') are also needed: these include manganese, copper, zinc, boron, sodium, molybdenum, vanadium, chlorine and cobalt.

Over most of the earth's surface the optimum conditions for growth do not occur. In the polar regions and above the snow line on mountain ranges the climate is too cold for plants to grow. Water is scarce in the deserts and growth there is restricted, or impossible. Most of the oceans have only a meagre supply of nutrients: they too, in their own way, are deserts. Much of the rest of the surface of the earth is deficient in some respect and the growth of living creatures is less than it might be. At the other extreme there may be an excess of some substance which, too, inhibits life. Pollution may be no more than too much of a good thing. Farming is essentially concerned with ensuring that, as far as possible, the optimum supply of nutrients and the right conditions for growth are provided for selected photosynthesizing organisms.

Within an organism photosynthesis occurs through a complicated series of chemical reactions. Substances such as chlorophyll, fructose, glucose and phosphoglyceric acid are formed and take part in the building of tissues. Some of the energy of the sunlight becomes locked into the chemical bonds of these substances and thereby stored. But most of the energy absorbed by the organism is used simply to maintain its activity or metabolism. A living creature has to use energy to take in nutrients, move them to where they are required, combine them into the substances it needs, control its temperature, adjust itself to light and shade and repair itself when damaged. On a summer day a beech tree, for example, may emit through its leaves five times their total weight of water and this must be replaced if the tree is not to wilt; the production of just 5 tonnes dry weight of a crop can require that some 2 000 tonnes of water are drawn from the soil.

Even under ideal growing conditions no more than a few per cent of the visible radiation absorbed by a photosynthesizing organism is used in the formation of organic tissue. The amount formed is called the gross photosynthetic production. It is produced during the daytime, with a release of oxygen. At night some of it is broken down, with a release of carbon dioxide, and the unlocked energy is used in the organism's self-maintaining activities. This is called 'respiration', a term usually denoting breathing, but now used by scientists as a general name for the breaking down of organic compounds in this manner within a living creature.

The rate of respiration varies widely between photosynthesizing organisms and depends on the conditions under which they are growing, and their maturity. A stable ecological system such as a mature rain forest, which is neither expanding nor contracting, will have a total respiration rate of 100 per cent. An intensively cultivated farm, on the other hand, will have a low rate of respiration and a high net production during the growing season.

The efficiency with which the biosphere converts solar energy to organic matter is in general very low. Under optimum conditions a couple of per cent of the visible radiation may be used, but more usually it is only a small fraction of this. The total production of organic matter in the whole biosphere is about 160 billion tonnes dry weight per annum: since each tonne has stored about 4 000 kilowatt-hours of energy, this represents a rate of utilization of visible radiation of about a fifth of a per cent. About a third of the photosynthetic production occurs in the seas and oceans; the rest occurs on land.

Photosynthesis is basically the same on land or in water. Without the proper combination of nutrients, climate and energy it cannot occur. But there are important differences. On land the nutrients are in the soil and are absorbed through the roots of plants. When plants die their constituents are returned to the soil and become available for new growth, provided they are not carried away by water flows, winds, grazing animals, or the intervention of man.

In deep water the situation is much more difficult for living creatures. Photosynthesis can only occur in the top layer of the water where there is enough energy from sunlight. In muddy water this might be only a few centimetres, in the oceans it could be a couple of hundred metres. But the nutrient supply is also crucial: when an organism exhausts the nutrients in its vicinity it must have the supply replenished, move to another area, or die.

In places where the ocean currents well upwards and bring sediments back to the surface the nutrient supply is replenished and marine life can thrive, as in the Humboldt current off the west coast of South America. On the continental shelves, too, there is a steady supply of nutrients brought down into the sea by the rivers. Cutting off this supply by, for instance, damming a river will reduce the amount of marine life. This happened in the Mediterranean off the coast of Egypt where the sardine catch was reduced after the construction of the Aswan Dam.

In deep, still water remote from land the amount of living matter which can be supported is very small. The sediments at the bottom may be rich in the required minerals but because there is no light, no photosynthesis can occur, and because there is no mixing of the upper and lower layers of water these vital supplies remain permanently inaccessible. The only way in which life can be supported is by the supply of nutrients blown by the wind, by that in the surface drift of water from richer areas, and that carried by other living creatures. The few organisms which are supported on this meagre supply grow and die and carry their share of nutrient to the depths where it remains. The organic productivity of the oceans averages about 0·1 grams dry weight per square metre per day; in comparison, a rich alluvial plain may average 200 times as much.

Solar energy, once fixed in organic tissue by photosynthesis, can be used by other creatures of the biosphere. The heterotrophic creatures which feed on it are called primary consumers: they are the grazing creatures, or herbivores. They include caterpillars, whales, cows, deer and those vegetarians who exclude eggs, milk and cheese from their diet. Secondary consumers, or carnivores, eat herbivores. They may in their turn be eaten by other carnivores. Such a succession of consumers is called a 'food chain'. Some creatures, of course, are not exclusively fixed in one position along a chain. Cats eat birds which feed upon seeds as well as those which feed on caterpillars: they will also tackle an owl which feeds upon small herbivores and carnivores. Man's diet has probably the widest range of all.

Most food chains are quite short, no more than three or four species, because of the very low efficiency with which energy is passed along the chain. Not all the organic material produced at any one stage can be used as food by the next creatures along the chain. Furthermore, just as the photosynthesizing plant or creature uses most of the energy it absorbs in self-maintenance, so do other creatures. A herbivore or carnivore has to move, breathe, chew, digest, excrete, regulate its temperature, mate and generally maintain itself. In the mature animal this consumes its full energy intake. In the growing animal energy is stored as the tissue builds up. As a very rough guide it can be taken that each step along a food chain involves an average energy loss of 90 per cent. The consequence is that the further along a chain a creature is positioned the lower its numbers must be.

The yield of grain from an area of land may be sufficient to feed, say, a hundred people. Assume that instead of grain the land is given over to grass production with exactly the same efficiency of solar-energy conversion as the grain. Cattle feed on the grass and convert, say, 10 per cent of it to flesh. Now the human population which can be supported is ten. If lions were imported to feed on the cattle and their flesh were the sole human food, the population sustainable would be just one person.

Of course this over-simplifies greatly. It ignores dietary considerations, the role of protein as opposed to that of carbohydrates, the use the body makes of fats, and so on. But the basic point is valid: the closer a creature is to the primary energy source, solar energy, the greater its numbers can be. This is why eagles and owls are so much rarer than the creatures on which they feed, and why the polar bear is such a lonely hunter. In the sea, with its longer food chains, the predators at the end of the chain are rare even by these standards.

The results of an analysis of the energy flows in an actual eco-system are shown in Table 1. These are taken from Howard T. Odum's famous study of Silver Springs, a stream in Florida with vegetation covering the bottom and numerous species of herbivores and carnivores. The net photosynthesis is only 0·5 per cent of the total insolation. Of this about 17 per cent is converted to flesh by the herbivores and about 4·5 per cent of this in turn is converted by the carnivores which prey on them. By the time the solar energy has

Table 1. Simplified flow of energy through a natural eco-system

	kilowatt-hours/metre²/year
Insolation	1 977·10
Absorbed by plants	476·83
Photosynthesized (gross production)	24·20
Photosynthesis less respiration (net production)	10·27
Intake by herbivores	3·92
Net production by herbivores	1·72
Intake by carnivores	0·45
Net production by carnivores	0·08

Source: Howard T. Odum, analysis of Silver Springs, Florida, quoted in David M. Gates, 'The Flow of Energy in the Biosphere', *Scientific American*, September 1971.

reached this last stage 99·996 per cent of the total entering the system has been reduced to waste heat.

Every creature dies. There is a range of creatures in the biosphere whose function is to process the complex compounds of dead organic matter and break them down into their simple original constituents, thus releasing the macro- and micro-nutrients which can then be used again by the photosynthesizers. These creatures are called the decomposers. They include the bacteria of decay, worms, insects, fungi, small animals, carrion eaters. Through their activity the cycle of the biosphere is closed, and in a stable ecosystem there is neither loss nor gain of material. The materials flow in a closed loop and this ensures the continuation of the system. If the loop is opened, the material is passed out of the system and unless it is replaced the system must collapse.

Human food production is one such opening of the loop. It removes nutrients from the land and rarely returns them. In this it is akin to mining the land. The food-production process can only be sustained by using fertilizers to replace the nutrients taken from the soil. The sun continues to supply the energy for photosynthesis but man has to provide the energy to produce and spread the fertilizers. When the American Forest Institute advertises that 'unlike coal, oil or ore – wood is renewable so we need never run out' it omits this important point. Wood can be used as an industrial raw material or a fuel but its continued cropping requires a return of all the nutrients extracted from the soil by each crop.

But this cyclic return of nutrients must not be confused with the energy flow. Energy is not recycled. It flows through the system, entering it as solar energy and leaving it as waste heat. The tiny fraction of the sun's energy entrapped by photosynthesis passes along the food chains of the biosphere. 'Food webs' might be a better term, because it brings out the complexity of the interrelationships between the consumers and the consumed. Each step in which the stored energy of one creature is converted into the motive energy and food of the next wastes a large proportion of the energy remaining in the food webs. When the final work of the decomposers has been done and the organic material has been reduced to its mineral constituents again the energy flow is complete. All the originally captured solar energy has been dissipated.

But throughout history there have been quirks of behaviour, slight malfunctionings of the system. At times some of the organic matter escaped the decomposers. It became locked into the sedimentary rocks of the earth's crust before it had been broken down. It turned into coal and oil and natural gas. In this way large amounts of solar energy and organic nutrients were stored in the world's coal and petroleum deposits.

It has been suggested that because he burrows deeper than any other creature, man has a unique role in the natural scheme of things. He is the only creature which can complete the work of decomposition of these ancient organic remains. Through his coal mines and oil wells he brings them to the surface, releases their energy into space and redistributes their constituents over the face of the earth.

2

Man in the Biosphere

The biosphere is sometimes portrayed as a region of harmony and peace. This is misleading. It is not a realm of friendly collaboration but of balance: the balance of opposing forces. The equilibrium is between the consumers and the consumed, both having a total common commitment to survival. Inefficiency or weakness or inability to adapt to changing conditions is punished by extinction. From the dinosaurs down, a multitude of species has been eliminated for the one crime the biosphere does not forgive: inability to stand the test of competition.

Man was originally just another creature involved in this unrelenting struggle, one of a group of related ape-like animals. He was a medium-sized omnivorous animal, not as common as a monkey but more common than a horse. During the early Stone Age (up to about 300 000 years ago) the human population was probably not more than about 100 000 and was confined to the African continent. Man lived as a hunter and gatherer of food, obtaining his body's energy requirements by foraging and scavenging, and by killing those creatures slow or careless enough to allow themselves to be caught. His limitations of size, speed, strength and agility confined him to a minor role in the biosphere.

The creature which is less efficient than another in collecting or digesting its food is making less than it might of the available energy. If times become harder it will be at a competitive disadvantage with another creature more efficient in utilizing the same food source and it will be displaced. Over the millions of years of evolution living creatures have therefore tended to reach an optimum efficiency in food collection and conversion for their particular position in the biosphere.

However, this optimum efficiency is a balance, not necessarily a maximum. Any creature so efficient in food gathering that it eliminated its food supply, could not survive more than one season. There must always be sufficient left over to produce next year's crop, or flock, of food. Nature's survivors are those which achieve the necessary balance. They are economical and efficient in the collection and utilization of their food; but they do not deplete the resources on which they depend.

Creatures generally become efficient by specialization. The birdlife of a wood illustrates the point. On the ground are the worm and snail eaters; treecreepers feed on the grubs and insects in the crevices of the bark; woodpeckers dig into the bark; seed and bud eaters are found at the outer ends of the branches; fly catchers hunt between the trees; predators and carrion eaters prowl through the whole system. This refinement of methods of food collection is countered by specialization in self-defence. The lower life forms ensure their continuance by a profligacy in seeding or procreation; the higher forms use camouflage, fleetness, protective armour, or well-equipped ferocity.

This development of capacities peculiar to itself defines the ecological niche which a creature makes its own. For some creatures this niche may be as rigidly defined as the stomach of a cow or the leaves of a particular tree; for others, such as a carnivore or carrion eater, the range will be much greater. But those physical characteristics which make a creature the most effective in pursuit of its own food usually bar it from making use of other foods, so that with increased specialization comes vulnerability to scarcities or change.

Man alone emerged from the niche to which his physical characteristics would have confined him. Where other creatures reacted to competitive pressures by increased specialization of their bodily forms and functions, man used his intelligence to make weapons and tools. A sharp stone, wielded by the hand, concentrates energy at its edge or point and can be used to split or bore. A stone-tipped spear makes a man as deadly as a lion or as swift as a cheetah. A woven net hung across a game path makes him a trap as subtle as that of a spider. A raft of logs lashed together gives him the hunting range of a crocodile or a water bird. As man's catch improved and his choice of prey was extended, so were his chances of survival and of increasing the species. And this was achieved without the need for specialized body characteristics. In developing these primitive devices man was outwitting the biosphere, managing to have his cake and eat it. The long tongue of the anteater makes it a very effective gatherer of ants but it prevents it eating nuts. The long spear of the early hunter enormously increased his food-collecting power but in no way reduced his versatility. For those who believe man should never have left his natural place in the harmony of the biosphere, the rot set in at least a million years ago when he started throwing stones.

The other crucial advantage man acquired over his fellow creatures was culture. This has been neatly defined as the 'non-genetic transmission of experience'. With its development man was no longer dependent for his advancement on the chance occurrence of favourable genetic mutations, which is the slow way the rest of nature adapts. For man the experience of one generation was immediately available to the next, and not only the knowledge but also an accumulation of capital in the form of weapons, tools and artefacts. If there is a natural order in which each creature has a place determined by its physical characteristics, man was perpetrating a gigantic fraud. By physical nature he was a land-based, slowish, weak, not particularly agile omnivore. But here he was rapidly extending his range and poaching into those areas hitherto reserved for the specialized animals of land, air and water.

Man also discovered new ways of using or diverting energy supplies to his own advantage by controlling fire, growing crops, and taming animals. He could use the vast stores of energy locked in the forests to keep himself warm and thus survive in harsh climates.

He could use flames to intimidate creatures stronger and fiercer than himself, and to drive his prey into traps. The seemingly limitless forests of the earth were a source of energy which he could mine for thousands of years.

Agriculture is generally thought to have begun in Asia Minor. The earliest traces of organized farming communities have been found in Asia Minor, in the 'Fertile Crescent' and are believed to date from about 8 000 years ago. By then man's impact on the biosphere was already noticeable: it is believed that by this time the sabre-toothed tiger and the woolly mammoth had been hunted to extinction. His increasing knowledge of the behaviour of the animals and plants of the biosphere and his well-developed social and culture grouping enabled him to embark on the organized cultivation of plants for food and the control of animals for a source of food and muscle power.

Agriculture is essentially a process by which the stored energy of photosynthesis is directed towards man. It begins with the selection and cultivation of plants. But there is much more man can do. The available solar energy is often wasted because there is not enough water or because the nutrient supply is deficient. Irrigation and the application of manure, fertilizers, or lime are remedies for these deficiencies, and by using them man opens wider the gate of photosynthesis through which solar energy becomes his food.

The selection of seeds for crops changes the characteristics of plants. Instead of adapting to enhance their own chance of survival, they are directed into evolutionary paths which serve man's interests: man's crops do not survive untended in the wild. In diverting water, in supplying fertilizers, in selective cultivation, man gains energy for himself. In general it is at the expense of creatures which would have thrived if the natural balance had been left undisturbed.

The domestication of animals also widens his control over the natural energy flows. Just as fire enabled him to tap the energy of the forests, the use of animals opened another storehouse of otherwise inaccessible energy. Every blade of grass converts solar energy, but is useless to man: he cannot eat it, nor is it a particularly good fuel. But if he puts cattle to graze on it, he can obtain animal milk or meat for food; animal hide for clothes or to cover a dwelling; animal bones or tusks as weapons or tools; animal dung to burn; animal muscle power to plough fields, haul loads, dig irrigation trenches, raise water. True, the efficiency of conversion of solar energy to flesh or work is low, but the area of grassland is large. With the domestication of animals the grasslands of the world became man's energy gatherers – vast natural solar collectors.

But all this energy was not entirely free. When he began organized cultivation and keeping domestic animals man also had to increase the amount of energy he spent in order to obtain useful energy in return. Grass had to be cut and stored as hay. The harvest cost human and animal energy. So also did the care of animals. And the same was true of slave labour. The use of human beings as domestic animals was a feature of the rise of early civilizations. The slave worked for a master and received no pay. But he consumed energy in the form of food. In this he was the same as an animal – though the problems of his maintenance and control were often more difficult. Both the slave and the domesticated animal were precursors of the mechanical engine; they were a means of turning the energy of food, or fuel, into the work their owners required of them.

By 2000 BC village and urban settlements were common and the pace of change was increasing. World population was reaching 100 million. With his growing ability to harness energy and provide himself with reliable food supplies man was finding himself with time to think and organize himself. Civilization and the founding of states and cities occurred when the increased efficiency of agriculture enabled society to exempt an increasing number of people from the daily toil of collecting food. They became the priests, craftsmen, politicians, thinkers, warriors, builders, designers. On the foundation of the energy supplied by the farming community they moulded dynasties in China and Babylon, planned pyramids and plotted the movements of the stars.

The waterwheel appeared around 100 BC in the West, though it was reputedly in use centuries earlier in China. The energy it tapped still came from the sun which had lifted the water in the mill-stream from the oceans, but the inefficiencies of biological energy conversion were eliminated. A moderate sized watermill could provide perhaps 10 kilowatts, an energy supply far beyond anything previously obtainable in such a concentrated form. The energy costs of constructing and maintaining the mill were trivial compared with those of feeding animals or human slaves. Once it was constructed the waterwheel was a source of free energy for years or decades. The windmill, known in the Arab world several centuries earlier, was brought to northern Europe in the eleventh century. It rapidly acquired a place beside the watermill as a major provider of energy in the medieval economy. The new availability of energy enabled the pace of social and technical development to quicken.

By the fifteenth century Europe had so advanced in control of its own environment that it could look beyond this and embark on voyages of 'discovery' to new parts of the earth whose people were 'primitive natives', scarcely to be regarded as human. When conflicts occurred the indigenous people were rapidly swept aside by the European superiority in weaponry, mobility and training in the abstract and practical arts of warfare. Backed by the big energy-consuming economies of their own countries, the explorers were more than a match for the hunter–gatherers, subsistence farmers and fishermen they encountered.

The superiority can be indicated in quantitative terms. An active western adult needs, on average, about 3 kilowatt-hours (2 600 kilocalories) of food energy a day. The average energy consumption per person in a late medieval community, counting that provided by fire, windmills, watermills and consumed by domestic animals, was about seven times as much as this. It was as though each citizen of such a community had at his disposal the work of six human slaves. Not only did this give him time to reflect, being free of much of the drudgery necessary for his sustenance, but he also had the power to put his thoughts into action. He could divert his surplus energy resources as he saw fit, and the community could construct palaces, cathedrals and universities, stage wars and circuses, and support musicians and philosophers – or conquer the New World.

Of course there was nothing equitable in the distribution of energy within the late medieval or feudal economies. Many people lived lives of real slavery and for the average man life was little better. The concentration of energy was in the hands of the minority. Energy and power are related in the technical sense that power is defined as the rate of energy flow but also in the sense that access to energy is access to political and physical power over others. The mill owner was often a tyrant in his community.

At the end of the eighteenth century the world's population was perhaps 700 million. But by then a crucially important division had appeared: that between the developed and underdeveloped countries. In the developed countries energy flows had been harnessed, population densities had risen, towns and cities had grown. Over the rest of the world most indigenous populations lived at the levels of energy consumption of hunter–gatherers or primitive farmers. The impact of these cultures on their environment was small. Perhaps they would, in their own time, have come along the same path as the Europeans; but they were never to be given that chance. The European way of life eliminated them, suppressed them, or absorbed them.

Nowhere at that time had European high-energy society reached a higher stage of development than in Britain. A truculent powerful country, it had pushed the capture of natural energy flows about as far as seemed possible. Its population was soaring. By 1800 it had exceeded 10 million, or about 44 persons per square kilometre. England on its own had a population of 8 million – a density of 66 persons per square kilometre. In comparison, the population density of a hunter–gatherer society is about 1 person per square kilometre and that of a primitive agricultural community about 10 persons per square kilometre.

Britain appeared grossly overpopulated for its resources. Its agricultural production seemed near its ultimate development and its forests were rapidly disappearing. Energy was being used a lot faster than it was being replaced. A limit seemed to have been reached.

3

The Industrial Revolution and the Oil Age

In 1798 Malthus published his famous essay 'On the Principle of Population'.[1]* Britain was in trouble: its population was rising rapidly and its agriculture showed few signs of being able to keep pace with the growing demand for food. The poor were on the verge of starvation.

Malthus thought the problem of growing and distributing enough food for the increasing population under the conditions then prevailing was impossible of solution. The land was still mainly controlled by the gentry. The cities were, for the majority of their inhabitants, filthy, overcrowded, disease-ridden agglomerations of human misery. He wrote that 'the perpetual tendency in the race of man to increase beyond the means of subsistence is one of the general laws of animated nature which we have no reason to expect will change'. And if humanity failed to recognize this fact he warned that 'we shall not only exhaust our strength in fruitless exertions and remain at as great a distance from the summit of our wishes, but we shall be perpetually crushed by the recoil of this rock of Sisyphus'.

One of Malthus' aims in his essay was to refute the optimism of Condorcet and Godwin. Writing after the French Revolution, they saw humanity proceeding steadily into a boundless era of peace, justice and plenty. Malthus' view of the future was a brutal and gloomy one. He used the paltry statistics available to him to support his case in a thoroughly modern way. He contrasted the compound, or exponential, growth of population with what he

*Superior figures refer to List of Sources Quoted, pp. 317–20.

believed was the simple, or linear, growth possible in agricultural production. The arithmetic is naïve; the conclusions drawn are at times blatantly tendentious. Nevertheless, the dark shadow of Malthus' vision of population outstripping the resources on which it depends has never been completely dispelled. That there are limits to resources is obvious; where these limits are is the question that remains unanswered.

To Malthus, famine in Britain was not a remote future possibility, it was an immediate threat. The recoil of the rock of Sisyphus was a danger requiring urgent action. In this he was to be proved completely wrong. He failed to see that in the workshops of the inventors, toolmakers and iron-wrights of the eighteenth century the way had already been prepared for the construction of an energy-consuming social and industrial system which would belittle man's earlier accomplishments and provide a means of escape from the closing trap of hunger.

In the eighteenth century, Britain had survived a resource crisis of a kind which had destroyed previous civilizations. There had been a serious danger in the early part of the century that lack of wood might cause a collapse in iron-making. The smelting of iron used large amounts of charcoal which was made from wood, and it was the presence of trees rather than iron deposits which determined the location of early iron-works. There were communities of iron-workers in Kent and the Weald of Sussex, but as the woods on which they depended were cut down they declined or disappeared. Without iron Britain would probably have reverted to a primitive agrarian society.

The country was saved by coal, for, in 1709, Abraham Darby discovered a method of smelting pig-iron using coke. Coal was readily available and had been used as a fuel for hundreds of years. It could be obtained from surface outcrops and sea-coal was quite common in some areas: one of the streets in the old City of London is still called Sea-Coal Lane. London even had an anti-pollution law published in 1273 forbidding the burning of coal in the city.

The threatened iron industry thus had a new source of fuel. But there were limits to the amount of coal which could be obtained in the traditional ways. It was a simple matter to dig coal from a surface outcrop and even to follow this some distance underground, but when the seam passed below the natural ground-water level, mining it depended on a continued pumping away of water – a task generally beyond human and animal muscle power. While Abraham Darby's discovery had averted the immediate crisis, new methods were required to increase coal supplies.

In 1708 Thomas Newcomen had built his first beam engine. This used the expansion of steam in a cylinder to drive a piston which was then sucked back by the partial vacuum which formed when the steam condensed. It was slow, cumbersome, dangerous and, by modern standards, hopelessly inefficient in its use of energy. But it provided the required solution to the problem of pumping water from underground mines. By 1765 there were a hundred of these machines at work in the Tyne and Wear coalfields. Later in the century the beam engine gave way to the more efficient rotary engine of Watt. The billions of tonnes of coal in Britain's coalfields were now accessible, and coal mining and iron working grew steadily throughout the eighteenth century.

Objects now taken for granted became commonly available for the first time. Chains, nails and iron hand-implements came from the small iron workshops. Scythes, sickles, and knives of increasingly high quality were produced by the growing cutlery industry in Sheffield and elsewhere. Iron replaced wood in the blades of ploughs and the bearings and hubs of cartwheels, thus increasing the effectiveness of draught animals. Marshlands were being drained and soils improved. By the time Malthus was writing, the country was ready for an unprecedented upward surge in agricultural and industrial outputs.

Hargreaves' spinning-jenny, Arkwright's spinning frame, Wedgwood's potteries, Roebuck's iron-works and ordnance factories: these had been the pioneering enterprises of the eighteenth century. In the nineteenth century a host of new inventors and entrepreneurs responded to the growing availability of energy and developed ways of using it – and obtaining more of it. Stephenson's locomotive introduced an era of prodigious railway building. Telford, Brunel and McAdam, with their associates and disciples, bursting with inventiveness and determination, constructed a network of roads, railways, viaducts and bridges which could carry food and manufactured goods to all parts of the country.

Science and technology advanced and the disciplines of civil, structural, mechanical, municipal and electrical engineering emerged. The navy became a powerful weapon in the creation of the greatest empire the world had yet seen. Merchant ships brought a flood of food and raw materials into the country and exported manufactured goods and coal. In 1900 coal exports were no less than 58 million tonnes.

In 1800 the total annual coal production of Britain was about 10 million tonnes. By 1850 this had climbed to 60 million tonnes and by 1900 it had reached 225 million tonnes a year. More coal was consumed in the two years 1899 and 1900 than in the whole century between 1700 and 1800. The country had come a long way from the agrarian, insular society of a hundred years before. Its population had grown from 11 million in 1800 to 38 million in 1900; Malthusian pessimism had not been vindicated by events.

At the beginning of the eighteenth century about 80 per cent of Britain's population was rural; by the end of that century the proportion had dropped to round 50 per cent, typical of an underdeveloped country today (see Table 2). By 1900 less than 10 per cent of the active population was engaged in agriculture. Thus, in complete contrast with all previous societies, most of the human and other energy of Victorian Britain was directed into trade, manufacture, building, learning, the arts, transport and so on; the proportion required to produce food was a tiny proportion of the total. For the majority of the population an almost complete divorce from the biosphere had taken place and food had become something obtained, not from the earth, but from a shop.

Coal made it possible for a society to feed itself on the produce of foreign farms. H. G. Wells wrote:

The sober Englishman at the close of the nineteenth century could sit at his breakfast table, decide between tea from Ceylon or coffee from Brazil, devour an egg from France with some Danish ham, or eat a New Zealand chop, wind up his breakfast with a West Indian banana, glance at the latest telegrams from all over the world, scrutinize the prices current of his geographically distributed investments in South Africa, Japan and Egypt and tell the two children he had begotten (in the place of his father's eight) that he thought the world had changed very little.[2]

Coal had made all this material progress possible. As a fuel for an industrial economy it had immense advantages over any other major energy source known at the time. The biosphere is slow in converting solar energy: a tree takes twenty or thirty years to grow to a usable size. The total dry-matter production of a temperate forest is about 11 tonnes per hectare per year, of which perhaps 50–70 per cent might be usable as fuel if cutting and gathering were carefully carried out. To harvest the equivalent of 225 million tonnes of coal in the form of timber would therefore require about 600 000 square kilometres of forest: the total area of England, Scotland and Wales is about 223 000 square kilometres. Ignoring all the practical difficulties of cutting and transporting the timber as well as the energy costs of forest maintenance and fertilizer inputs, Britain was simply not big enough to produce the energy it was consuming by any cultivation of the biosphere.

Coal supplied in a concentrated, portable form the energy of hundreds of thousands of square kilometres of forest. With a productivity of over 200 tonnes per man per annum in 1900 each miner was producing the energy equivalent of about sixty hectares of forest. The comparison with other energy sources is equally devastating. A reasonably large sail windmill is capable of producing about 30 kilowatts under favourable conditions and might work for 2 500 hours in a year. Over thirty such windmills would be required to produce the equivalent energy-output of a single coal miner.

During the nineteenth century the rate of growth in energy con-

Table 2. Percentage of active population employed in agriculture – selected countries, 1850, 1900, 1950

Country	about 1850	about 1900	about 1950
AFRICA:			
Algeria	—	—	81
Egypt	—	70	65
South Africa	—	60	33
AMERICA:			
Canada	—	42	20
USA	65	38	13
ASIA:			
Japan	—	71	48
India	—	—	74
EUROPE:			
Belgium	50	27	12
France	52	42	30
Germany	—	35	24
Great Britain	22	9	5
Ireland	48	45	40
Netherlands	44	31	20
Spain	70	68	50
Sweden	65	54	21
USSR	90	85	56

Source: Quoted in Carlo M. Cipolla, *The Economic History of World Population*, Penguin, 1970.

sumption in Britain exceeded the rate of population growth by a factor of nearly five. Coal consumption per head in 1800 was under a tonne a year; by 1900 it had grown to nearly four and a half tonnes. The average Victorian (if there was one) had at his service the energy equivalent of a slave retinue of thirty-five people.

Karl Marx noted that the main beneficiaries of the increased energy flow were the bourgeois. He wrote indignantly that

> the extraordinary productiveness of modern industry, accompanied as it is by a more extensive exploitation of labour power in all other spheres of production, allows of the unproductive employment of a larger and

larger part of the working class, and the consequent reproduction on a constantly extending scale, of the ancient domestic slaves under the name of a 'servant class', including men-servants, women-servants, lackeys etc.[3]

And certainly the sweating kitchen staff of the large houses, enslaved winter and summer by those gigantic coal-devouring ranges, would have been hard put to see the liberatory effects of so much energy consumption. The copious availability of energy had made the Industrial Revolution and the development of Victorian Britain possible, but it had not determined the precise path of social development. The newly created wealth could have been distributed in many different and better ways.

The Industrial Revolution was not, of course, confined to Britain. But it came later elsewhere. America and Europe avoided some of Britain's worst horrors and were able to benefit from the technological progress already made. The early locomotives and steam engines had efficiencies of no more than a few per cent – if they were well made and working properly. Later developments were much more efficient. In other words they provided more useful work for the same consumption of energy. European and American industry was not handicapped by accumulated equipment of low efficiency, and their pace of development could therefore be more rapid. Industry around the great coalfields of the Ruhr and the Saar, of Belgium and Alsace-Lorraine grew quickly and began to compete very effectively indeed with that of Britain. The British iron and steel industry, for instance, which in 1850 dominated the world, had only a 10 per cent share of world production by the beginning of the First World War.

By then yet another energy revolution was well under way. Like coal, oil had been known and used for thousands of years. Seepages at ground level were common enough to make it a reasonably well-known commodity – but in quantities more appropriate to gourds and flasks than super-tankers. It was sometimes encountered in drilling for brine, and as the brine well had then to be abandoned, oil was regarded as an unmitigated nuisance. An account of one such occurrence in Kentucky in 1829, however, shows that it could at least provide good spectator sport. The brine well

> vomited forth many barrels of pure oil . . . 25 to 30 feet above the rock. . . . About two miles below the point at which it touched the river it was set on fire by a boy and the sight was grand beyond description. An old gentleman who witnessed it says he has seen several cities on fire but that he never beheld anything like the flames which rose from the bosom of the Cumberland to touch the very clouds.[4]

Commercial drilling for oil began in 1859. Edwin Drake, an ex-railway conductor turned entrepreneur, with the self-bestowed rank of 'Colonel', used a percussion rig to drill a bore hole in Pennsylvania which struck oil at a depth of 69 feet. Drake earned his place in history, though he died in poverty on a small pension provided by the oil industry. The oil boom had begun and within a year 175 further wells had been drilled in the same area.

Coal was superior to wood as an energy source and oil in its turn was superior to coal. Writing in 1915 in Britain, Professor Herbert Stanley Jevons, a leading authority on the coal industry, discussed the advantages of oil. It was, he said, more easily stored and handled; it was cleaner; its use eliminated dirty and dangerous jobs such as those of stokers on ships; and having a 50 per cent higher calorific value than coal, it enabled the paying cargo of ships to be increased. It could be used 'in the invention of Dr Rudolph Diesel whose internal combustion engine contained several new and ingenious features', and which was 'three and a half to four times as efficient as the steam engine'.[5] Professor Jevons was right, and there were yet more advantages, the most important being the ease with which oil could be obtained. Spindletop, the famous Texas oilwell, blew out as a 'gusher' in 1901, at a rate of over a hundred thousand barrels a day. This was the energy equivalent of the work of 37 000 miners, simply running to waste. Oil was much cheaper than water: at Spindletop oil cost 3 cents a barrel and water 5 cents a glass.

Production and consumption of oil rose rapidly and by the turn of the century American oil production had reached 9 million tons a year. When Jevons was writing it had reached about 40 million tons a year. No one believed that production could continue at this rate, let alone continue to expand. Jevons wrote of the 'scare writings of people of vivid imagination who see in recent engineering progress "the dawn of the oil age"' and reassured his coal-industry readers that 'the extensive general adoption of mineral oil as a power producer for all purposes is a very unlikely contingency for, as far as our present knowledge extends, the supply of oil is strictly limited' and the United States which produced most of the world's oil had 'probably reached its zenith' of production. Here he was completely wrong.

Despite the incredulity of both experts and non-experts, new and rich oilfields were discovered in increasing numbers. Production doubled every eight or nine years. Apart from a brief period in the early 1900s when Russian production was actually greater, America remained the largest producer. Table 3 shows oil-production statistics for different areas of the world from 1890 onwards. Up to the 1950s America was still producing, and consuming, well over half the world's oil. While other countries still retained the distinctive characteristics they had evolved during the coal-based Industrial Revolution, America was leading the way in the development of the social and industrial patterns of the oil-based high-energy society.

The most obvious feature of this society is its mobility. Only in the wild imaginings of fairy tales about magic carpets and witches' broomsticks had mankind ever before liberated itself from the tyranny of time and space. For most of his existence the pace of man's movement had been limited to five or six miles an hour and that for short distances. The bicycle, of course, had been a vehicle of social revolution, liberating at least some of the Victorians – as *The Complete Cyclist* said in 1897, '... women, even young girls, ride alone or attended only by some casual man friend for miles together through deserted country roads.'[6] The train too had been a miracle in its own time. But now instant mobility came within the

Table 3. Growth of world oil output – 1890–1978

Date	East Asia	Austral-asia	Middle East	Africa	E. Europe USSR	West Europe	North America	South America	WORLD
					tonnes × 10⁶				
1890	—	—	—	—	3·5	—	6·4	—	9·9
1900	0·1	—	—	—	9·6	—	10·1	—	19·3
1910	2·8	—	—	—	12·1	0·2	29·2	0·2	44·5
1920	4·1	—	1·7	0·1	6·1	0·1	82·2	1·0	95·2
1930	7·8	—	6·4	0·3	23·6	0·2	128·6	26·0	192·9
1940	11·2	—	14·1	0·9	36·9	1·5	192·6	36·8	293·9
1950	13·6	—	87·8	2·3	42·1	2·7	284·3	88·2	519·3
1960	31·1	—	263·5	14·4	163·0	13·5	392·3	173·5	1 051·2
1970	80·7	8·9	706·0	302·5	372·3	15·7	563·3	237·0	2 286·5
1973*		*152·3*	1 047·0	290·5	438·1	22·6	586·2†	245·1	2 781·8†
1974*	—	*164·3*	1 077·5	272·5	468·3	22·6	563·9†	221·9	2 791·0†
1977*	—	*230·0*	1 116·5	305·2	560·0	70·1	538·8	184·9	3 005·5
1978*	—	*236·4*	1 054·1	297·1	593·0	89·7	572·5	185·5	3 028·3

Source: E.N. Tiratsoo, *Oilfields of the World*, Scientific Press, 1973.

*These data are taken from the *B.P. Statistical Reviews of the World Oil Industry*. They are not strictly comparable with those in Tiratsoo and do not give a breakdown for 'East Asia' and Australasia. The figures in italics are for South-East Asia, South Asia, China and 'Other Eastern Hemisphere' combined.

†Excludes natural-gas liquids production in the US.

reach of everyone; the Ford Model T annihilated distance. America created the first mobile society.

American cities, designed around the car and for the car, began to spread across the countryside over distances impossible to cover on foot. Superhighways were built. Patterns of social behaviour, shopping, entertainment, family life and courtship were fashioned around the new mobility and thus created a continuing and growing need for it. By the eve of the Second World War there were nearly 30 million cars in the United States, one for every four people in the country. But the energy cost was high. By 1940 the average energy consumption in the United States was approaching twice that in Britain at the height of its imperial grandeur. The energy slave retinue of a US citizen had increased to sixty or seventy.

From the beginning of the present century the United States has, in fact, been glutted with energy. Up to the late 1960s the oil industry's constant problem was the country's potential for overproduction which kept the market weak and the price of oil and petrol extremely low. During the 1930s a system of production-regulation called 'proration' was created. Under its rules owners of oilwells were allowed to produce only what the market would bear. Allowed production was often down to 2 per cent, or even less, of the installed productivity capacity of some oilfields. One writer[4] has referred to the introduction of widespread proration as the 'beginning of the conservation era', but the conservation was of price-level only. There was no question of advocating restrictions in use. On the contrary, since the problem was a surfeit of oil, efforts were constantly directed to finding ways of increasing consumption.

In this Americans were extremely successful. The huge private car, produced by the million in the factories of Detroit, consumed more energy than a whole Roman legion, and was used to take its owner down to the drugstore for a packet of cigarettes. Central heating, air-conditioning, skyscrapers, houses crammed with mechanical and electrical devices, a huge consumption of food, clothes and material goods, in all these ways, and more, American society devised ways of consuming energy undreamt of previously. All of them could only occur in a society which had energy in generous excess over that needed for the basic necessities of life.

America also led the way in the automation of industry. This is usually thought to be a way of making industry more efficient: in fact it substitutes mechanical for human energy. It is a way of using the energy of coal, oil, gas or some other fuel to do work which would otherwise have to be done by hand. While this increases the total productivity per worker it generally raises the total energy cost of production also. Workers in automated industry have higher outputs than their counterparts elsewhere simply because they have a greater number of energy-using machines to help them.

Another feature of the early American development of the high-energy society was the waste of natural gas often found in conjunction with oil. Natural gas is an almost perfect fuel and even more valuable as a feedstock for the chemical and fertilizer industries. But because the pipelines necessary for its distribution took a long time to lay and the oil companies usually would not wait to get a return on the money they had invested in drilling the oilwells, whenever gas was discovered it was flared or allowed to blow away. The waste of fuel was prodigious: in total it was the equivalent of hundreds of millions of tonnes of coal. When the oil industry got round to using it, natural gas became so popular as a fuel that in 1973 it was supplying almost a third of the United States' total energy requirements and was rapidly moving into scarce supply. The years of waste were bitterly regretted. The fact that the international oil industry has continued with the same policy of wasting gas in other areas where there is no immediate market for it will no doubt be a cause for future regrets as well.

The United States characterizes in an extreme form the high-energy society. Table 4 shows its energy consumption in 1978 in comparison with a number of other countries: with less than 6 per cent of the world's population, it consumes almost 30 per cent of the world's energy.* Its standards of consumption are so high that they can never be equalled by more than a very small proportion of humanity.

As a country industrializes, the proportion of its population engaged in food production decreases, as can be seen in Table 2. In the early years of industrialization the major effort of the economy goes into the production of goods. But as the industrial society matures the emphasis begins to shift towards the provision of services. An increasing proportion of the population is employed in finance, insurance, sales, repairs, professional activity, and government. This has been described as the shift towards a

*The US is frequently castigated for this disproportionate share of the world's total energy consumption. It is little known that in the past the imbalance was much worse. In 1955 the US consumed 40 per cent of the world's energy and in 1947 the share was 47 per cent. Joel Darmstadter quoted in 'Exploring Energy Choices'.[7]

Table 4. World energy consumption – selected countries, 1978

Country/Area	Oil	Natural gas	Coal	Water power	Nuclear energy	TOTAL
	tonnes of oil equivalent $\times 10^6$					
USA	887·9	504·2	355·0	79·3	75·5	1 901·9
Canada	86·9	47·3	19·2	56·0	8·5	217·9
Latin America	202·0	42·3	15·2	46·6	0·7	306·8
Western Europe						
Austria	12·0	4·5	1·6	5·8	—	23·9
Belgium & Luxembourg	28·4	10·5	9·4	0·1	2·7	51·1
Denmark	16·0	—	4·0	—	—	20·0
Finland	12·5	0·8	3·7	2·4	0·8	20·2
France	119·0	20·9	28·2	15·0	6·4	189·5
Greece	12·1	—	0·3	0·6	—	13·0
Iceland	0·6	—	—	1·3	—	1·9
Republic of Ireland	6·0	—	0·6	0·2	—	6·8
Italy	99·5	24·4	10·2	12·2	1·1	147·4
Netherlands	37·2	34·0	3·0	—	0·9	75·1
Norway	8·7	—	0·5	20·9	—	30·1
Portugal	7·3	—	0·5	2·8	—	10·6
Spain	47·0	1·5	10·7	10·7	2·0	71·9
Sweden	26·6	—	1·0	14·9	6·1	48·6
Switzerland	13·4	0·7	0·3	8·4	2·1	24·9
Turkey	15·2	—	4·5	2·3	—	22·0
UK	94·0	37·9	70·4	1·2	7·9	211·4
West Germany	142·7	42·1	48·0	3·6	8·2	244·6
Yugoslavia	14·9	1·6	1·5	6·8	—	24·8
Cyprus/Gibraltar/Malta	1·5	—	†	—	—	1·5
Total Western Europe	**714·6**	**178·9**	**198·4**	**109·2**	**38·2**	**1 239.3**
Middle East	83·3	30·1	†	1·0	—	114·4
Africa	60·3	8·3	49·2	12·0	—	129·8
South Asia	37·1	9·7	67·2	10·7	0·6	125·3
South East Asia	105·6	7·9	47·2	4·1	1·3	166·1
Japan	262·6	17·0	54·0	17·4	12·5	363·5
Australasia	37·7	7·8	20·9	8·2	—	74·6
USSR	412·8	289·0	340·0	46·0	11·0	1 098·8
Eastern Europe	100·4	65·0	265·0	5·3	4·0	439·7
China	84·7	33·0	380·0	8·0	—	505·7
Total World	**3 075·9**	**1 240·5**	**1 811·3**	**403·8**	**152·3**	**6 683·8**

Source: *BP Statistical Review of the World Oil Industry*, 1978.

'post industrial society'.[8] In America 1956 can be taken as a date of transition. In that year for the first time, 'the number of white-collar workers (professional, managerial, office and sales personnel) outnumbered the blue-collar workers (craftsmen, semi-skilled operatives, and labourers)'.[8] Another feature of the 'post-industrial society' is the amount of time which must be spent in education in order to be able to obtain, and retain, employment within the ever more complicated industrial and social system with its increasing rate of obsolescence of skills and technologies. The 'learning-force' – that is all those at school or engaged in further training – outnumbered those working in the United States for the first time in 1965.[9] Since these activities are generally less energy-intensive than manufacturing, the later stages of economic development will tend to require less energy per unit of economic growth than at the beginning.

The pattern of development of the United States is, however, uniquely its own. It created its own distinctive social and industrial forms around very high energy consumption and heavy dependence on oil, and, in recent decades, natural gas. In the older industrial countries of Europe energy consumption is lower and the use of oil came much later than in the United States. In 1950, for instance, British oil consumption was just 13 million tonnes, or a quarter of a tonne per head, whereas in the United States average consumption was then 4 tonnes per head.

This, in outline, is the rise of the high-energy society. Man's development since his emergence as a hunter–gatherer can be traced through his growing ability to control the energy flows of the biosphere; and to draw on the accumulated capital of coal and oil. By 1900 the prototypical coal-based industrial economies were firmly established. But since then the dominant position of coal has been steadily eroded. Table 5 shows how its percentage share of world energy supplies has fallen from over 94 per cent of the total of around 900 million tonnes of coal equivalent in 1900 to around 30 per cent at present.

This relative decline of coal, however, has not meant that less of it is being consumed. As Table 6 shows, its consumption has almost doubled since the end of the 1930s. But in the same time oil consumption has increased nearly thirteen times, amply justifying the designation of this present phase of human development as the 'oil age'.

Table 5. Percentage of world energy supplied by main fuels – 1900–1978

Fuel	1900	1920	1940	1960	1965	1978*
Coal/lignite	94·2	96·7	74·6	52·1	43·2	27·1
Oil	3·8	9·5	17·9	31·2	36·7	46·0
Natural gas	1·5	1·9	4·6	14·6	17·8	18·6
Hydro-electric	0·5	2·0	2·9	2·1	2·2	6·0
Nuclear	—	—	—	—	—	2·3

Source: Quoted in E.N. Tiratsoo, *Oilfields of the World*, Scientific Press, 1973.

*1978 Data from *BP Statistical Review of the World Oil Industry*, 1978.

Table 6. World energy consumption – 1929–76

Date	Solid fuel	Oil	Natural gas	Hydro/ nuclear	Total	per capita kg of coal equiv./ year
			tonnes of coal equivalent $\times 10^6$			
1929	1 367	255	76	14	1 713	867
1937	1 361	328	115	22	1 826	900
1950	1 569	636	273	41	2 519	1 054
1955	1 816	948	397	59	3 211	1 200
1960	2 204	1 323	620	86	4 233	1 403
1965	2 250	1 919	926	118	5 213	1 588
1970	2 388	2 855	1 421	157	6 821	1 893
1974	2 524	3 524	1 561	207	7 816	2 017
1976	2 696	3 733	1 662	228	8 318	2 069

Source: *UN Statistical Year Books*

Note: Data in this series are not strictly comparable with those shown in Table 4. For details of the different conversion factors used by BP and the UN, see Appendix. Note particularly the different conventions for hydro/nuclear.

4

The Unifying Principle

Early Greek philosophy anticipated a number of modern scientific theories in its search for a unifying natural principle. Democritus proposed a theory of the atomic nature of matter in about 400 BC, and a hundred years or so before that Heraclitus had asserted that the universe was in a constant state of flux and that its unifying principle was fire. But these promising speculations remained undeveloped and were later obscured by the development of the Platonic and Aristotelian cosmologies.

The modern theoretical view of the universe, in which the concept of energy can be used to link the whole range of observed phenomena, has been built on the work of Isaac Newton. This extraordinary man dominated scientific thought from the latter half of the seventeenth century through to the end of the nineteenth. His work is still the basis of practical engineering dynamics and machine design. His Laws of Motion made it possible, for the first time, to quantify the interaction of physical objects and opened the way to an understanding of the fact that energy is neither created nor destroyed.

In Newtonian, or classical, mechanics, energy is defined as the capacity to do work – work is said to be done when an object is moved or altered. 'Work' in common usage is useful activity but the technical definition is indiscriminate and says nothing about the usefulness of the action. A man sweeping leaves into a pile is doing work; but so is the wind in scattering them again. In the British system of units which was developed during the early Industrial Revolution, and is still used in some parts of the world, the unit of work is the foot-pound: the amount of work done when a weight of one pound is raised a distance of one foot.

A weight of one pound raised a distance of one foot can be said to 'possess' an energy of one foot-pound. Placed on a balance or lever, it is able, in descending, to raise another weight of one pound through a distance of one foot. In this way an energy transfer takes place: the first weight 'gives' its energy to the second. Objects which, because of their position, are capable of doing work on other bodies are said to possess 'potential' energy.

Anything which is moving can be made to do work in coming to rest. By measuring the amount of work objects do in being brought to rest it is possible to establish the relationship between speed, mass* and energy. It is found that the energy of a moving body, its 'kinetic' energy, is directly proportional to its mass and proportional to the square of its speed. A moving two-tonne car will therefore have twice the energy of a one-tonne car moving at the same speed. But doubling the speed of a vehicle quadruples its energy: a crash at 100 kilometres per hour involves four times the energy of a crash at 50 kilometres per hour.

With these concepts of work and energy, it is possible to unify a wide range of phenomena. A flowing river, a tightened bowstring, a spring wound in a clock, water stored behind a dam, and a horse, to take some random examples, are all capable of doing work. This work can be measured and expressed in precise quantitative units, such as foot-pounds.

It was also obvious that the energy of motion was closely related to heat and that a theoretical account of work was incomplete without a statement of this relationship. The friction of two unlubricated metal surfaces could produce enough heat to fuse them.

*The use of the word mass in everyday language is quite different from its scientific use. Mass can be thought of as a measure of the amount of matter in a body. Weight on the other hand is the gravitational force acting on a mass. On the moon, for example, the weight of a body is a sixth that on earth. Weightlessness can be survived, but anyone experiencing 'masslessness' is beyond recall.

Large quantities of water were needed to take away the heat produced in the boring of a cannon. There was plenty of evidence for the connection between heat and work but the credit for formalizing it into a scientific proposition goes to James Prescott Joule, an English brewery owner who spent his life in scientific research. He determined the mechanical equivalent of heat in a classic experiment in 1843. In this experiment, a heavily insulated container of water with a rotating paddle inside it is connected by a string to a weight over a pulley. The weight is allowed to move downward, causing the paddle to rotate and thus agitating the water. At the end of the experiment when the weight has reached the bottom of its travel and the water is still again, it is found that the temperature of the water has gone up. From this Joule was able to measure the amount of heat produced by a known amount of work. He found that it took 772·5 foot-pounds to raise the temperature of 1 pound of water through 1°F. Accurate modern measurements have corrected this figure to 778·16 foot-pounds, which is within 1 per cent of Joule's value and a tribute to his experimental skills. This quantity of energy is known as Joule's mechanical equivalent of heat. It is also called a British thermal unit (Btu).

But the concept of energy could be extended far beyond this. In 1831 Faraday had shown that an electric current could be generated by moving a magnet near a conducting wire, thus relating mechanical work and electrical energy. And Joule himself had shown that the rate of heat production when a current flowed in a wire was proportional to the square of the current times the resistance of the wire. James Clerk Maxwell's electromagnetic theory of light, which was one of the major achievements of the late nineteenth century, subsequently brought together the theories of electricity, magnetism and light in a simple mathematical formulation. In the words of Professor J. D. Bernal: 'The electro-magnetic theory was a crowning achievement which realized the dream of Faraday that all the forces of Nature should be shown to be related . . .'[10]

Now there was a common factor, energy, quantifiable in all the phenomena with which science was concerned, from the invisible waves of electromagnetic radiation to the motion of the planets. Even the metabolism of living creatures could be measured within the same system. Priestley had discovered oxygen in 1774 and shown how it was 'consumed' in burning – and in breathing. Later, Lavoisier was able to quantify the metabolic behaviour of living creatures, showing that they behaved just like a fire, 'burning' the materials they absorb as food by combining them with oxygen, and releasing their energy in the form of heat.

As the nineteenth century reached its end it seemed that the frontiers of science had finally been delineated; only the details remained to be filled in. The Newtonian concept of the universe had been vindicated. Science rested securely on the twin principles of the conservation of matter and the conservation of energy. The quest for the philosopher's stone which would transmute the base elements into gold had long been abandoned. The elements had been identified and allocated their fixed place, in accordance with their properties, in the scheme of the universe. It was fundamental to the theory of chemistry that at the end of a chemical reaction it should be possible to account for all the elements which had entered into it. In other words matter was neither created nor destroyed. Similarly energy was neither created nor lost. In an ideal closed system the total quantity of energy before and after an event was the same. Its distribution between different parts of the system might change and it might appear in different forms, but the total amount could not vary.

Then in 1895 Röntgen discovered X-rays. He found that, when he was using a cathode tube, invisible rays capable of penetrating solid objects were escaping from the apparatus. The excitement of the discovery led to a rush of experiments elsewhere. Within months Becquerel had discovered the radioactivity of uranium. The Curies went on to discover radium and describe its properties. Rutherford, working at McGill University in Montreal with the chemist Soddy, did pioneering work in elucidating the nature of the mysterious phenomenon of radioactivity. From the fever of scientific investigation two conclusions emerged which undermined the existing foundations of science: elements were changing from one into another, and energy was apparently being generated spontaneously out of nothing.

Einstein later resolved many of the paradoxes by showing that an even more comprehensive synthesis of existing scientific theories was possible. According to his special theory of relativity, mass and energy were equivalent. The relationship could be expressed in quantitative terms: $E=mc^2$, energy is equal to mass multiplied by the velocity of light squared. The release of energy observed in radioactivity was the result of a genuine transmutation of the elements and a conversion of some of their mass into energy. With this Einstein was able to transcend the Newtonian concept of the physical universe and make it part of a wider understanding of the nature of physical reality. Table 7 gives some impression of the unifying power of the concept of energy. Units commonly used in the separate disciplines of electricity, atomic physics, dietetics and engineering can not only be expressed in terms of each other, but all can be expressed in units of mass.

Table 7. Equivalents of various energy units

UNIT	joule	electronvolt	kilocalorie	kilowatt-hour	kilogram mass
joule	1	$6{\cdot}242\times10^{18}$	$2{\cdot}389\times10^{-4}$	$2{\cdot}778\times10^{-7}$	$1{\cdot}113\times10^{-17}$
electron-volt	$1{\cdot}602\times10^{-19}$	1	$3{\cdot}828\times10^{-23}$	$4{\cdot}450\times10^{-26}$	$1{\cdot}783\times10^{-36}$
kilocalorie	$4{\cdot}186\times10^{3}$	$2{\cdot}613\times10^{22}$	1	$1{\cdot}163\times10^{-3}$	$4{\cdot}658\times10^{-14}$
kilowatt-hour	$3{\cdot}600\times10^{6}$	$2{\cdot}247\times10^{25}$	$8{\cdot}598\times10^{2}$	1	$4{\cdot}007\times10^{-11}$
kilogram mass	$8{\cdot}988\times10^{16}$	$5{\cdot}610\times10^{35}$	$2{\cdot}147\times10^{13}$	$2{\cdot}497\times10^{10}$	1

The proliferation of energy units is, in fact, a cause of considerable inconvenience. An attempt has been made to simplify scientific measurements by devising a common system of units for all the scientific disciplines, the Système International d'Unités (SI), which, it is hoped, will be eventually adopted throughout the world. This system was initiated at the Ninth General Conference of

Weights and Measures in 1948. Its standard units of mass and length are the kilogram and metre. Temperature is measured in degrees Celsius (°C) or degrees Centigrade. Absolute temperatures are measured in kelvin (K) which are measured from absolute zero which is taken as −273·15°C.* The freezing point of pure water is thus 0°C or 273·15K at atmospheric pressure.

The concept of 'power' deserves special mention because it is so often misunderstood. In scientific language power is the *rate* at which energy is expended or work is done. When James Watt was selling his steam engines he described their capacity for work in familiar terms. A 1-horsepower engine was capable of working at the same rate as a good horse. He defined this as 33 000 foot-pounds per minute which was the rate at which he found a horse could raise a weight suspended over a pulley. The *amount* of work done is the rate multiplied by the time. In one hour the standard horse, therefore, does 1 980 000 foot-pounds of work or 1 horse-power-hour.

The SI unit of power is the watt, a very much smaller unit than the horsepower (1 watt=0·00134 horsepower). The unit of energy derived from it is the joule, a rate of work of 1 watt maintained for 1 second (not to be confused with Joule's mechanical equivalent of heat). This is also a very small unit and the kilojoule (1 000 joules) or megajoule (1 000 000 joules) are more commonly used. A mega-joule is about 0·37 of a horsepower-hour. The kilowatt-hour (kWh), however, is much more familiar: it is a rate of work of a thousand watts maintained for an hour. The kilowatt-hour is the 'unit' of electricity, the amount used in an hour by the normal single-bar electric fire. It is about 1⅓ horsepower-hours.

The First Law of Thermodynamics states that energy is neither created nor destroyed. It is therefore incorrect to say that energy is 'consumed'. When it passes through an energy-using system it is merely *transformed*. When a power station burns coal to generate electricity all the energy released can be accounted for: in the cooling water, the exhaust gases released up the chimney, the heat losses from the machines and the transmission wires, the heat

*The temperature at which an ideal gas has zero volume and pressure.

emitted from the machines and appliances at the point of use, and any energy stored in the final product – the electricity – could, for instance, be used to charge a battery, or bring a kettle of water to the boil. None of the energy is lost or destroyed.

But there is a sense in which energy is 'consumed'. A great number of energy-using processes are not reversible. The universe is, in fact, running down. The sun is getting older and colder; the stars are burning out. All the energy can be accounted for but not recaptured for re-use. Whenever energy is used some of it is lost and flows towards the 'heat-sink' of outer space. There it adds infinitesimally to the random movement of atoms more thinly dispersed than any vacuum science has yet managed to create. This is the end of energy: the ultimate graveyard from which no calorie or electronvolt can ever return. Lord Kelvin was the first to talk of the inevitable 'heat death' of the universe when all its energy has become uniformly distributed as low-grade heat.

The Second Law of Thermodynamics, formulated by Clausius in about 1850, states: 'No process is possible whose sole effect is the removal of heat from a reservoir at one temperature and the absorption of an equal quantity of heat by a reservoir at a higher temperature.' In other words, heat does not move spontaneously from a colder object to a warmer one. Expressed in such careful language the Second Law seems almost a truism, but in its application it is of fundamental importance to scientific understanding. All energy-using processes are subject to this limitation. A pot of boiling water can be used to warm a bathtub of cold water, but a bathtub of warm water will not bring the pot to the boil.

'Efficiency' is defined as the amount of useful energy produced by a system as a proportion of the total energy input. It can be shown that the efficiency of a 'heat engine', that is any mechanical system using heat energy, depends on the temperature drop which occurs within the system.* The practical result of this is that,

*For a theoretical heat engine, the efficiency is given by the formula $(T_1-T_2)/T_1$, where T_1 is the initial temperature of the working fluid and T_2 its final temperature, expressed in degrees K.

mechanical improvements such as better bearings apart, the only way of improving the efficiency of heat engines is by increasing their operating temperatures. In modern power stations the steam temperatures are close to the limits tolerable for machines made of reasonably available materials and the *theoretical* efficiencies are around 65–70 per cent, with operating efficiencies of about 40 per cent.

It also follows that the efficiency of low-temperature energy systems will be very small. The low efficiency of solar energy conversion in the biosphere is not a result of nature's ineptness. Plants convert solar energy in a process which operates over a temperature drop of just a few degrees. It is the nature of the universe that only a small proportion of the energy entering such a system can be converted to a more useful form.

An important practical conclusion to be drawn from this view of energy is that of 'thermodynamic matching'. For maximum efficiency of energy use, 'high-grade' energy should only be used where necessary. It is thermodynamically absurd, for instance, to use hydrocarbon fuels to produce electricity at an efficiency of 30 per cent and then use the electricity to produce low-grade heat for space heating. If the hydrocarbon fuels were used directly they could produce the heat required at perhaps 70–80 per cent efficiency. The convenience of electricity is dearly bought in wasted energy. Another approach to energy use is called 'thermodynamic cascading'. This places energy uses in a descending thermodynamic order such that the energy 'waste' emerging from each use is of sufficiently high grade to meet some or all of the energy requirements of the next use.

Clausius also introduced the term 'entropy', now defined as the measure of the degradation of a system's thermal energy and hence its unavailability for conversion to mechanical work. Thermodynamics is a notoriously difficult subject to understand, and entropy is one of its most elusive concepts. An outline understanding of the basic principles of thermodynamics is, however, important. Popular misconceptions can lead to unjustified optimism about the potential of resources and technologies. In particular, the limitations imposed by the Second Law need to be clearly appreciated. Not all energy is equally usable. When the energy resources of the earth are being counted, not only the quantity of energy, but its 'quality', too, needs to be considered.

5

'Social Energetics'

In the early years of the twentieth century the continued economic expansion of western industrial society was taken for granted. Its financial systems were demonstrably effective in rewarding initiative and progress, and eliminating the inefficient. Technical advances and discoveries were steadily expanding industry's productive capacities. The behaviour of the economy could be described, and predicted, by the classic market theories. In these, only prices were to be questioned; supplies of energy and raw materials were assumed to be automatically and indefinitely available to anyone with the money to pay for them.

One person who questioned the whole basis of industrial society, as it was then understood, was Frederick Soddy. This brilliant, but difficult man, who was awarded the Nobel prize in 1921, had collaborated with Rutherford, and later identified isotopes and coined their name. In the years immediately before the First World War he began to put forward the idea that it was necessary to look further than the laws of economics for a true understanding of society. He recognized the connection between economic activity and energy consumption. He saw the crucial importance of the change from an economy based on the use of constantly renewed, or replaceable, energy resources to one which used coal. Instead of relying on his energy revenue, man had begun to spend the accumulated capital of hundreds of millions of years of sunshine. The economic chaos which followed the war convinced Soddy that society was deluding itself dangerously.

He wrote in 1922:

The laws of energy under which men live furnish an intellectual foundation for sociology and economics, and make crystal clear some of the chief causes of failure not only our own but, I think also, of every preceding great civilization. They do not give the whole truth, but, in so far as they are correct to physics and chemistry, they cannot possibly be false.[11]

In a slightly later work he said of human history that

progress in the material sphere appeared not so much as a successive mastery over the materials employed for making weapons – as the succession of ages of stone, bronze and iron honoured by tradition – but rather as a successive mastery over the sources of energy in Nature and their subjugation to meet the requirements of life.[12]

He is also well worth quoting for the clarity of his description of the relationship between the flow of energy and materials and the activity of social and economic systems:

A continuous stream of fresh energy is necessary for the continuous working of any working system, whether animate or inanimate. Life is cyclic as regards the materials consumed and the same materials are used over and over again in metabolism. But as regards energy it is unidirectional and no continuous cyclic use of energy is even conceivable. If we have available energy we may maintain life and produce every material requisite necessary. That is why the flow of energy should be the primary concern of economics.[12]

More clearly than anyone else of the time, Soddy saw the precarious foundation on which the achievements of industrial society rested. His suggestion that energy should become the primary concern of economics was farsighted, but he did not apply himself to it and failed to expand it coherently. His work appears to have been one of the fascinating dead-ends of history. H. G. Wells based his fantasy *The World Set Free* on Soddy's work on radioactivity but none of his contemporaries seems to have attempted to develop his ideas on the integration of energy and economics into a consistent social theory.

In the early 1970s interest in relating social patterns to the flow of energy through society was again awakened, this time by the work of the American ecologist, Howard T. Odum. He and his brother Eugene Odum have specialized in analysing the flows of energy through natural ecological systems. This painstaking and meticulous work requires a measurement of all the energy entering and leaving a system as well as a chart of the many pathways it follows within the system. The well-known study of Silver Springs in Florida by Howard Odum was referred to in Chapter 1, in the discussion of the energy losses which occur along a natural food chain.

Odum has devised a notation in which various energy functions – an energy source, energy store, control, energy loss and so on – are represented by symbols and connected according to the energy flows. Figure 3 shows a typical representation of an aquatic eco-system using this notation. At the extreme left are the energy sources: light and in-flowing organic matter. Large plants and phytoplankton transform the light energy into plant tissue. This tissue enters the food web which supports a variety of herbivorous and carnivorous forms of life. Below each is shown the symbol for a heat sink, representing the unavoidable loss of usable energy in each transformation.

Howard Odum believes the same basic analysis can be applied to the analogous eco-system of human society. Man is a creature of the biosphere and ultimately depends, like all other creatures, on photosynthesized solar energy for his food. He imports energy in the form of fossil fuels into his eco-system and taps the energy of wind and water, and uses these in his activities of manufacturing, building, transport and farming. All these are energy-transformation processes subject to the laws of thermodynamics. They can be analysed and represented in the same way as the energy transformations of a natural eco-system.

Energy diagrams, for example, can be used to show how industrial society obtains its food. Figure 4 shows, in a generalized way, how this happens. At the left is the symbol for energy inputs representing all the energy, solar, human and fuel, used in agriculture. Then

Figure 3: Energy flows through a marine estuary eco-system

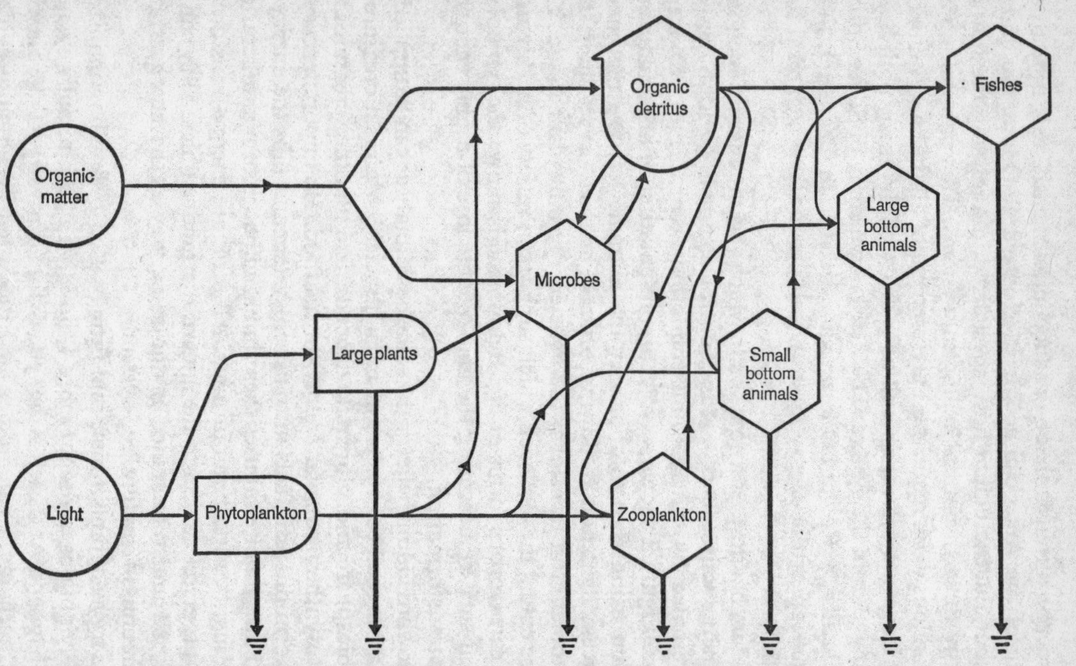

Source: Howard T. Odum, *Environment, Power and Society*, Wiley-Interscience, New York, 1971, Chapter 3.

comes the photosynthetic production of crops. These in turn are fed to food processors: cattle, chickens, millers or market operators. The produce moves through the food web, each operation involving an energy loss, shown by the heat sink, until in the end it reaches man.

Odum examines a variety of systems in this way and demonstrates the fallaciousness of some arguments which do not take energetics sufficiently into account. In monsoon farming, for instance, an animal may consume more food than appears economically reasonable. But the animal has the function of acting as an energy store for the farmer. Because of this energy store the farmer can start food cultivation immediately the monsoon conditions are right. Energy then 'flows' from the animal into the cultivation process. This energy subsidy, applied at the right time, enables the food crop to outdistance its weed competitors and exclude them from the tilled land.

A similar energetic analysis of the role of the sacred cow in India suggests that the case for eliminating them is less strong than might appear at first sight. The cows are not simply parasites. They glean food which would otherwise not enter the human food web and make it available as milk; the bullocks are necessary for farming work, and dung is the most widely used fuel. They are, like any other energy processor, an energy drain, but they can have a critically important function within the total energy system. Odum claims that an energy analysis of traditional religious practices would show many of them to be based on programmes for the control of energy flows and this is certainly borne out by, for instance, Rappaport's study of the primitive Maring people of New Guinea in his book *Pigs for the Ancestors*. When circumstances change, however, the continued observance of out-dated rituals may be counter to the interests of society.

As a way of preparing for the future Odum believes it is necessary to try to understand the whole of society in energy terms. History, the government of the Roman empire, even the heaven, hell and earth of religion, can be shown in the form of his energy diagrams.

Figure 4. Energy flows through the human food-production system in an industrial economy

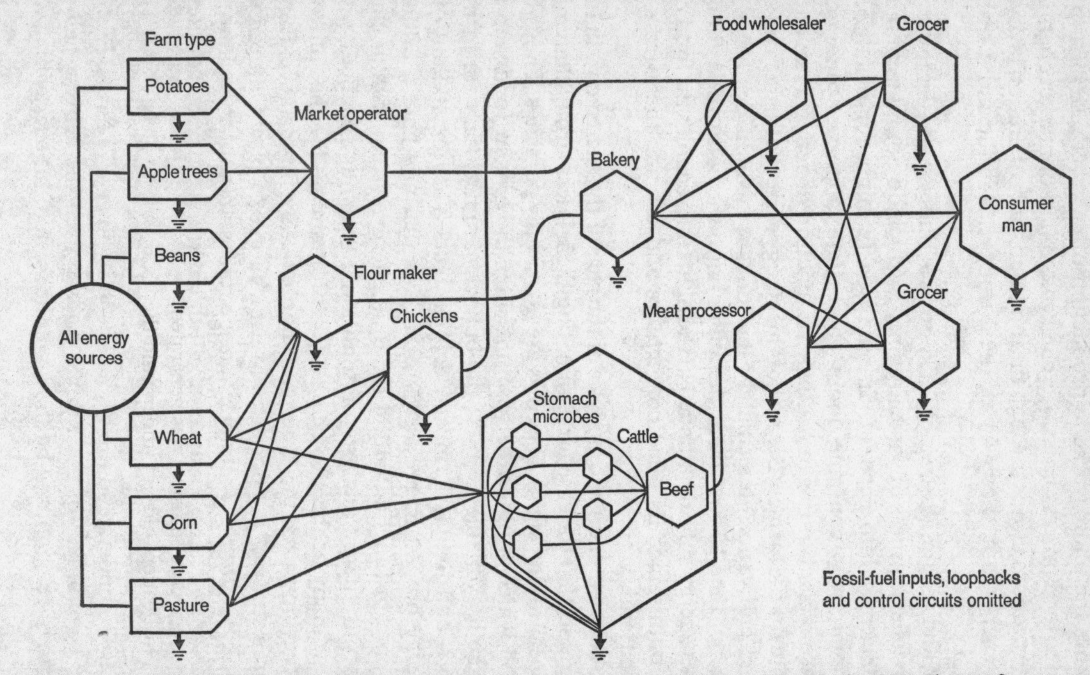

Source: Howard T. Odum, *Environment, Power and Society*, Wiley-Interscience, New York, 1971, Chapter 3.

In his major work *Environment, Power and Society* he is quite explicit about his aim:

... energy language is used to consider the pressing problem of survival in our time – the partnership of man in nature. An effort is made to show that energy analysis can help answer many of the questions of economics, law and religion already stated in other languages.[13]

He sees the control of energy flows as being the basis of all social organization and warns against the illusion that industrial society is permanent or self-sustaining. He says: 'Most people think that man has progressed in the modern industrial era because his knowledge and ingenuity have no limits – a dangerous partial truth. All progress is due to special power subsidies, and progress evaporates whenever and wherever they are removed.'[13]

The production of food, or any other commodity, requires both direct and indirect energy inputs. Soddy seems to have been the first to see this clearly. Discussing the energy costs of a car he looks at the tyres and says:

... If we pursue the tyres to their origin we shall find out how much of their cost is due to expenditure of energy. They call for a flow of the solar energy of a particular climate, physical labour in the rubber plantations, coal for the railways and ships that transport the raw materials from the tropics, as well as for the factories where it is made into tyres. These railways and ships, again, and all the buildings and equipment necessary for their manufacture, no less than the materials they use – the iron and metals and the coal which have to be mined – are the results of the expenditure of physical energy.[12]

Odum takes the analysis further. He is scathing about the optimists who believe that man has learned how to out-produce nature with the techniques of modern farming. Of this he says:

A very cruel illusion was generated because the citizen, his teachers, and his leaders did not understand the energetics involved ... A whole generation of citizens thought that ... higher efficiencies in using the energy of the sun had arrived. This is a sad hoax, for industrial man no longer eats potatoes made from solar energy; now he eats potatoes partly made of oil.[13]

Modern agriculture undoubtedly produces more food than primitive agriculture, but at a price – and a risk. The fat cow in the battery farm and the scrawny beast picking scraps from between the stones in an arid field are biologically similar and their basic rates of metabolism cannot differ by much. As energy-convertors their efficiencies must, therefore, be similar. In a free-ranging state, however, the cow has to devote a high proportion of its energy intake to the tasks of keeping warm or cool, finding food, fighting off the attacks of predatory animals, insects, microbes and viruses, and finding a mate. And in each generation large numbers of its population must be sacrificed in the endless evolutionary struggle to develop protection against predators and disease. The species evolved will therefore be tough, not at all what is wanted for the table. The trick in battery rearing is to remove all energy-consuming tasks from the animal and to develop genetic characteristics tending towards succulency. The result will not be a survivor if the going becomes rough.

The battery cow is then free to turn a great deal more of its food into flesh. It does not mean that the other jobs necessary for its survival do not need to be done. They are performed instead by machines and people: the atmosphere of the battery farm is controlled by heating and ventilating equipment; machines bring in food and remove waste; teams of vets, scientists, genetic engineers and technologists devise ways of optimizing usable flesh production and fighting off disease; tractors, harvesters and lorries sow, reap and transport the fodder. The cow plays the part of a beef machine in the elaborate process of energy conversion. The fact that the ancillary-energy consumption figures do not appear in the usual analysis of the efficiency of beef production can seriously distort the true picture.

There is no such thing as a free energy source. The utilization of any energy resource by man requires an expenditure of energy. When a resource is being evaluated the most important figure, therefore, is the net energy gain obtainable from it. If the only way of catching fish were to use chopped-up fish as bait it would be obvious that the process could only be justified as long as it did not require a subsidy of fish from some other source to keep going. Odum makes this point:

> Oil and coal will not run out, but the ratio of energy found to energy spent in obtaining them will continue to increase until costs exceed yields. If the net yield of potential energy begins to approach that of wood we will have returned to the solar-energy-based economy and by that time the standards of living of the world will have retrogressed to those of two centuries ago. Whether such changes will come suddenly in a catastrophe or slowly as a gradual trend is one of the great issues of our time.[13]

Soddy was concerned that classical economics had missed the point that the fossil-fuel resources on which society depends are both finite and depleting. Odum restates this even more strongly. Their objection is still valid. Economics has not yet devised a satisfactory way of dealing with resource depletion. Instead, it concerns itself with the mechanisms of resource distribution. It deals with relative scarcities. It can only presume that somewhere an adequate substitute exists for every resource on which humanity now depends. This is surely absurd. The market is not endowed with creative powers and there is no reason why some essential resources should not become exhausted.

Some people have entertained the superficially attractive idea of using 'energy units' instead of currencies. One of the earliest writers to consider this was H. G. Wells, who wrote in *The World Set Free*: 'Ultimately the government . . . fixed a certain number of units of energy as the value of a gold sovereign . . . and undertook, under various qualifications and conditions, to deliver energy upon demand as payment for every sovereign presented.'[2]

The supposed advantage of linking currency with energy is that the energy units are physically measurable quantities which do not vary between countries, are not subject to inflation or devaluation and cannot be forged, or printed by governments. With an energy-based currency the 'true' value of objects could be calculated in terms of the energy consumed in their manufacture and distribution. National and international trade, instead of being based on arbitrarily valued pieces of paper, could be related to the intrinsic energy value of goods and services.

A variation on this theme is the use of an 'energy ration'. Here is how one writer envisages this happening: 'Each government will from time to time establish the energy available to its national economy, and upon this basis allocate to each citizen exactly a year's supply of energy coupons . . . Every service and all goods will carry not only a price tag, but an energy tag.'[14] The idea is that people would have to consider the energy costs of what they purchased. This would encourage efficiency in energy use and therefore begin to optimize economic activity in terms of energy.

Although these are attractive ideas in some ways, their implementation would involve so many practical difficulties that it is unlikely it will ever occur. Energy units, it is true, are definable physical constants, but there is more to the definition of energy than its quantity. A kilowatt-hour of electricity is much more useful in that form than when it has been converted to the heat-energy of a bathful of warm water. An energy unit of currency, or an energy-ration coupon, would therefore need to specify not only the quantity of energy but the form in which it was available. This would mean energy would have to be specified as, say, barrels of oil of a particular calorific value. Arguments would undoubtedly begin to proliferate as the exchange rates of 'standard' oil with coal, natural gas and uranium were worked out. Technical changes would lead to the need to revalue one form of energy against the others. The impossibility of shipping great quantities of oil round every time there were trade transactions would lead to the use of notes promising to 'pay the bearer on demand' a certain number of barrels of standard oil. Some countries would have the temerity to consider leaving 'the oil standard' to give themselves freedom to develop their economies in their own way. The difficulties are familiar.

6

Energy Correlations and Costs

The general connection between economic activity and energy consumption began to attract an increasing amount of attention from economists and energy planners in the late 1960s and early 1970s. This was mainly because of the need to produce long-range energy demand forecasts as an aid to making investment decisions in the electricity supply industry.

This industry had seen an almost uninterrupted technical advance over the previous two decades. Part of the price of this progress, however, was that power stations had got very large. They took a long time to plan and build and they cost a great deal of money. This brought inflexibility into energy planning and a need to predict energy demand far into the future if shortfalls in supply or excess power station generating capacity were to be avoided.

The apparent correlation between energy and economic activity promised to provide the predictive method required. Figure 5 shows GDP* and per capita energy consumption for a large number of countries. The broad relationship is quite clearly identifiable. The higher the energy consumption the higher the GDP. The US with the highest energy consumption has the highest per capita

*Gross domestic product (GDP) and Gross national product (GNP) seem to be used almost indiscriminately in this context. In UK official statistics GNP is defined as the total value of goods and services produced, together with net property income from abroad. GDP omits the property income from abroad. The difference is small – about 1½ per cent in the case of the UK.

Other definitions which differ slightly in some technical respects are sometimes used, but the differences are well within the margins of error of the correlations discussed here.

Figure 5. Per capita GDP and energy consumption – selected countries, 1972

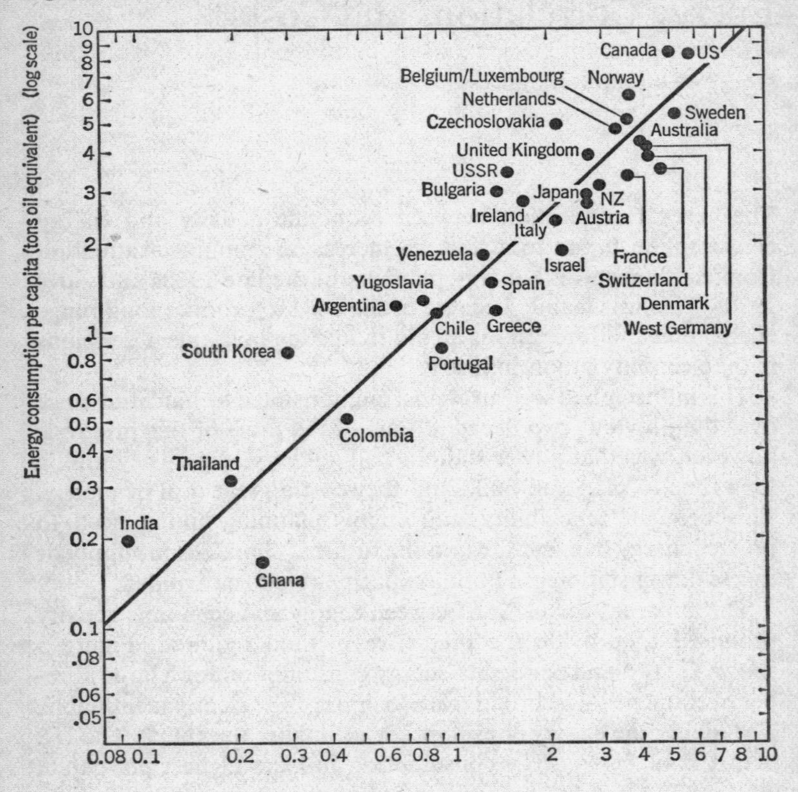

Source: J. Darmstadter, J. Dunkerly and S. Alterman, *How Industrial Societies Use Energy*, Johns Hopkins University Press, 1977.

GDP. At the other end of the graph in countries such as India, where much of the population is near subsistence level, the average energy consumption and per capita GDP are a minute fraction of those in the rich industrialized countries. It should be noted, however, that this does not tell quite the full story; in these countries a considerable amount of non-commercial energy in the form of fuelwood and dung is consumed and this does not appear in the energy-consumption figures.

A close examination of Figure 5 reveals many more anomalies. Although Sweden has almost the same per capita GDP as the United States it consumes only 60 per cent of the per capita energy of the United States. The USSR and Switzerland on the other hand have roughly the same average energy consumption, but the per capita GDP of Switzerland is three times that of the USSR.

The relationship between energy consumption and economic activity is far from simple and cannot be relied upon as a basis for predicting energy demand. The energy efficiency with which GDP is produced is affected by factors which include changing patterns of industrial activity, differences in the shares of total energy consumption provided by different fuels, and a host of other possible influences.

Work has been done in an attempt to find ways of ironing out some of these differences between countries so that their energy consumption can be compared on an equal basis. Two American economists, Adams and Miovic, tried to establish a series of 'efficiency coefficients' for different fuels – oil is more easy to use efficiently than coal, for instance – which could be used to make allowance for the difference between fuel use in different countries. At times the professional debate has been extremely acrimonious. The argument has, however, been virtually ended by the publication of the study, *How Industrial Societies Use Energy*.[15] This picks apart the energy consumption of nine major industrial countries. It shows that their energy consumption and GDP production are so enmeshed in the geographical, industrial, economic, social and political characteristics of countries that useful comparisons are almost impossible to make. The authors say: 'No tidy reckoning is possible on these questions. Our study points to complex and diverse reasons for intercountry differences in energy consumption.

Variations in energy/output ratios should not in themselves be viewed as indicators either of economic efficiency or even of energy efficiency . . . in numerous of its aspects, energy consumption is essentially a by-product or, at best, only one element within the wider framework of societal arrangements and choices.'[15]

The relationship between energy consumption and economic growth has also been examined to see if it remains constant over time within an individual country. It would be extremely convenient for energy planning if it did; many people wish it would; and some have managed to convince themselves that it does. The most common approach is to look at what is generally called the 'energy coefficient'. This is the relationship between the increase in GDP and the increase in total energy consumption in a particular year. An energy coefficient of 1·0 means that if the GDP increases by, say, 3 per cent there will also be a 3 per cent increase in energy consumption.

Just as the broad trend in the international comparison of countries is clear, so also is that of the energy coefficient within most countries. As GDP has risen so also has energy consumption. But closer examination begins to reveal the anomalies. In any particular year economic growth may or may not be associated with a proportional increase in energy consumption. In the twenty years between 1953 and 1973 the UK energy coefficient varied between +4·1 and −0·4, with the long-term average tending towards about 0·6. In other mature industrial countries there have been similar fluctuations in the energy coefficient and a rather higher long-term average of about 0·8.

Events since 1973 have, however, shown little respect for historical precedents. Throughout the OECD countries the average energy coefficient for the 1973–7 period fell to 0·4; in the European Community it was −0·4; in the UK for the whole decade 1969–78 the average energy coefficient has only been 0·3. The conclusion is that energy consumption and economic growth may be linked or they may not. Producing more goods at exactly the same average level of energy use as today's will require proportionally more energy. But there is no reason to presuppose that any particular increase in GDP will take place at exactly the same energy intensity as in the past. A higher output in the watch or electronics industry will, for example, require less energy than an increase in output of the same value from the iron and steel industry.

There are also examples of possible changes which have no effect on GDP but which alter energy consumption. Improving the heating and ventilation controls in a factory or office block produces the same GDP with less energy. To take a very much more homely example: hanging the curtains in front of the radiator beneath a window rather than behind it is an extremely common practice in some countries. Positioning the curtains behind the radiators so that more of the heat is directed out into the room rather than out through the window has no effect on GDP* but it saves energy and so reduces the energy coefficient.

There are other effects which can distort the energy coefficient. Technical change within industry in general works to increase the efficiency with which energy is used to produce goods and materials; the efficiency of electricity generation has doubled in the past thirty years, to take just one example. Substituting imported steel for that produced domestically has a much greater effect on energy consumption than it has on GDP. Increasing the insulation of houses or lowering the temperature of public buildings reduces energy consumption.

When all is considered, the puzzle is not so much that the energy coefficient deviates from a supposed norm but that anyone should imagine it should have a constant or predictable value. The apparent historical link between energy consumption and economic growth is one of accident rather than causality. What has happened in the past is a guide to the future only if a decision is made to repeat it in all its aspects. It is by now clear that the trends of the 1950s and 1960s cannot be repeated. Looking back at the energy coefficients to which they give rise provides little useful guidance on how to manage the future.

*The contribution of the energy industries to GDP is fairly small – about 5 per cent in the UK – so that energy saving will not in itself reduce GDP significantly.

GDP and total energy consumption are, of course, abstractions. They are highly generalized, or aggregated, figures. They do not reveal anything about how energy is actually consumed. Nor do they give information on how patterns of energy use change with time. To discover about these it is necessary to examine energy consumption in more detail.

The flow of energy through an industrial economy is exceedingly complex. Figure 6 shows diagrammatically what happens in the UK. As can be seen, the primary-fuel producers use some of their own energy to produce energy; they also draw on the output of the secondary-fuel producers. The term 'secondary-fuel producers' is used to describe those industries whose function is to take primary fuels and process them into a more useful form. The main secondary-fuel producer is the electricity industry. The others are usually taken to be the gas, coke and smokeless fuel producers.*

The secondary fuel producers use the output of the primary fuel producers; but they also feed back into their own processes some of their own output. The electricity industry obviously has to use some of its own electricity to heat and light its generating stations and offices; and the coal industry uses some of the electricity produced by burning coal to produce more coal.

Next comes energy distribution. Whether it is done by lorries carrying petroleum products or electricity wires, further losses occur. The path followed by a country's energy on its way to its final users is anything but direct; it is also costly in energy. As Figure 6 shows, almost 45 per cent of the UK's total energy went to the secondary-fuel producers to be converted to a more convenient form. The energy cost of the conversion was 570 618 million kilowatt hours. This was almost a quarter of the country's total primary-energy consumption. By far the greater proportion of this energy loss occurred in the electricity industry. Of the total

*The petroleum-refining industry could well be classified as a secondary-fuel industry. Conventionally, however, it is regarded as a primary-fuel industry and its energy consumption, as that of the coal industry, is listed under that consumed by 'primary-fuel producers'.

fuel consumption of over a thousand billion kilowatt hours in power stations, no less than 68 per cent was completely lost. Put another way, the country spent over 21 per cent of its total energy on the convenience of having energy available in the form of electricity. This energy expenditure was almost as much as the total energy consumption of the whole of British manufacturing industry.

Table 8. Energy consumption by final users: 'heat supplied' basis – UK, 1960 and 1976

	1960 kilowatt-hours	1976 (kWh) × 10⁶
Agriculture	17 609	20 278
Iron and steel	207 796	183 335
Other industries	417 789	488 337
Railways	83 739	13 889
Road transport	134 370	290 836
Water transport	16 701	15 278
Air transport	23 381	51 945
Domestic	422 653	431 115
Public services and miscellaneous	156 520	181 668
TOTAL	1 480 558	1 676 681

Sources: *UK Digest of Energy Statistics*; G. Leach, C. Lewis, F. Romig, A. van Buren and G. Foley, *A Low Energy Strategy for the United Kingdom*, IIED, London, 1979.

Table 8 gives a breakdown of the energy consumption by the main consumers in the UK in 1960 and 1976. These figures are calculated on what is called a 'heat supplied' basis. This means that just the calorific value of the energy supplied is counted. No distinction is made between a kilowatt hour in the form of coal and one in the form of electricity; neither is any account taken of the energy losses in conversion and distribution. While the picture is much more detailed and informative than that provided by the aggregated total energy consumption, it still omits many important elements.

Figure 6. Distribution of energy consumption – UK, 1972

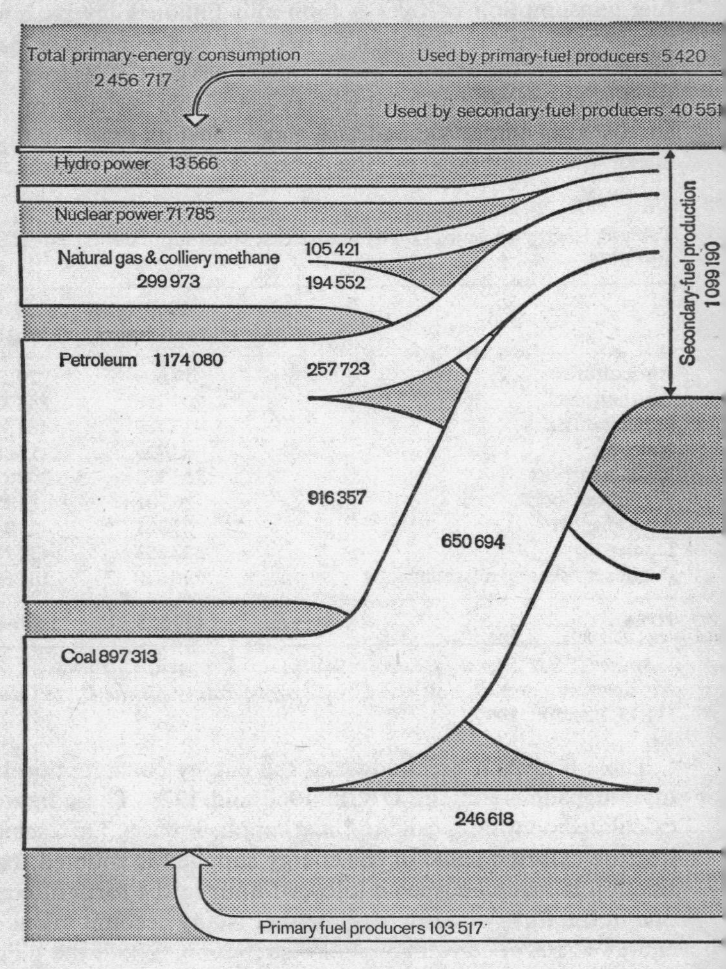

Source: *UK Digest of Energy Statistics*, 1973. (Diagram from Architectur
Association thesis by John Seed.)

Exports 3 340

Losses in conversion & distribution 600 240

ke, gas, etc. 58 395

ctricity 541 845

Coke & other
uels 164 871

own gas 94 141

lectricity 205 510

Net consumption by final users 1 700 894

66 892

147 174

Other uses 214 066

2696

Transport 354 120

351 424

186 494

Domestic 421 598

235 103

mary fuels
plied directly
inal users 1 254 010

193 673

Industry 711 111

517 438

Units: kilowatt-hours x 10^6

Nevertheless, the comparison between 1960 and 1976 reveals a number of interesting points. During that time of great change the total energy consumption on a heat supplied basis, increased by only 13 per cent, less than 1 per cent per year. The increase in primary energy consumption was, however, much greater: it rose from 265 million tonnes of coal equivalent in 1960 to 330 million tonnes of coal equivalent in 1976, a rise of almost 25 per cent. This difference is accounted for mainly by the increased production of electricity, which used 64·5 million tonnes of coal equivalent in 1960 compared with 114 in 1976. The rise in total energy consumption of the 1960s is therefore not a reflection of large increases at the point of use, but rather of a shift towards providing an increasing proportion of the country's delivered energy in the form of electricity. This proportion increased from about 6·5 per cent of the delivered energy in 1960 to about 13 per cent in 1976. The future proportion of electricity in the total delivered energy is thus a crucial element in determining the total primary energy requirement; so also, of course, is the split between electricity provided by coal or oil-fired power stations and those relying on hydro or nuclear power.

Further points reveal themselves with a detailed examination of the figures in Table 8. Road transport more than doubled its energy consumption; it increased its share of the total from 9 per cent to over 17 per cent. It and air transport were by far the fastest growing sectors of energy use. Road transport alone accounted for 80 per cent of the total increase in delivered energy during the period; most of this increase was because of the extraordinary growth in car traffic. The number of cars in the UK increased from 5½ million in 1960 to over 14 million in 1976. On the railways, freight traffic fell by about 23 per cent and passenger traffic by 12½ per cent, but energy consumption was cut to a sixth of what it was in 1960. This remarkable saving was achieved by eliminating the appallingly inefficient, though picturesque, steam locomotive.

Domestic energy consumption increased by only a very small amount. This was despite an increase of almost a quarter in the number of dwellings and a very evident increase in the standards of comfort and convenience. Undoubtedly a major reason for this was the big decline in the use of the open coal fire as a means of heating; almost any other method of heating is more efficient than this. From supplying nearly three-quarters of the country's domestic heating in 1960, coal fell to providing just a fifth in 1976. The balance was taken up by increases in the use of natural gas, electricity and oil.

An interesting way of taking the analysis of energy flows further is to measure not just the direct energy consumption of particular activities but to include the indirect consumption as well. Take the example of manufacturing industry: in addition to the energy it uses directly as fuel it 'consumes' the energy already used in the production of the raw materials, components, machinery and buildings it uses.

Total energy costs can be calculated by using what are called input–output tables which are produced by the statistical services of some governments. These tables work by dividing the economy into a large number of sectors and listing the amount each sector consumes of the output of all the others. Thus it is possible to discover from the UK tables how much of the output of the machine-tool, iron-and-steel, textile, rubber, footwear and energy industries is consumed by the car industry. By making an initial approximation based on the direct energy consumption of each sector and progressively refining it in an iterative series of calculations, the total direct and indirect energy consumption of each sector can be found. The UK tables list 90 separate divisions; those in the US use no less than 363. With such large numbers of sectors the calculations involved are extremely elaborate. Even with the use of large computers the results take years to obtain and tend to be of historic rather than contemporary interest.

In Table 9 the 1963 energy costs of various commodities, manufactured goods and activities in the US are listed. In the case of industries with a reasonably homogeneous output, such as coffee or cigarettes, the energy costs can be expressed as kilowatt-hours per

Table 9. Energy costs of producing goods and services – US, 1963

Commodity	Energy cost kWh/kg unless otherwise stated	Commodity	Energy cost kWh/1963 $
Cotton	15·7	Guided missiles	7·4
Tobacco	26·4	Ice-cream	17·0
Iron ore	0·504	Canned fruits and veg.	20·0
Butter	27·4	Bakery products	15·8
Cheese	20·3	Logging camps	23·6
Milk	7·35	Paper	47·2
Flour	2·36	Newspapers	14·2
Sugar	5·74	Business forms	15·9
Coffee	41·9	Greeting cards	12·6
Cigarettes	0·037/cig.	Concrete blocks	30·1
Paper	17·02	Farm machinery	22·9
Synthetic rubber	35·8	Sewing machines	13·1
Reclaimed rubber	6·10	Hotels	9·7
Steel	3·27	Hospitals	9·2
Copper	22·1	Non-profit organization	10·0
Lead	7·14	Fertilizers	36·3
Zinc	10·5	Macaroni and spaghetti	16·5
Aluminium	24·4		

Source: D. J. Wright, 'Goods and Services: An Input–Output Analysis', *Energy Policy*, December 1974.

kilogram or per unit of output. In the case of bakery products or sewing machines this cannot reasonably be done; and it is impossible to do it in the case of hospitals or non-profit organizations. For these a different approach is required. The output of the bakery industry or that of sewing machines can be measured in terms of the dollar value of the final product. In the case of hospitals or non-profit organizations the 'output' is usually measured by the salaries paid to employees or, where appropriate, the expenditure by consumers on the services offered. Thus the energy costs in these sectors must be expressed in kilowatt-hours per dollar. Further problems are caused by the need to produce consistent methods of dealing with price changes, inflation and fluctuations in the rate of international exchange; these alter the monetary value of output without affecting the energy consumption and could therefore appear to make processes more or less energy-efficient. There is also the problem of dealing with imported materials and components – should their energy content be counted or not? The method is obviously very cumbersome and the results are difficult to interpret with any certainty. Certainly they cannot be used to rank processes in a merit order, or to provide guidance on which activities should be encouraged. The manufacture of ice-cream is two and a quarter times as energy-intensive as that of guided missiles: it is an interesting fact but hardly one from which any firm conclusions can be drawn.

Interesting energy analyses have been done on the energy costs of agriculture. Here the solar energy input is ignored; only the energy costs of capturing it are counted. Moreover, human energy is such a small proportion of most processes in an industrial economy* that it can be ignored without noticeable effect on the final result. In the case of subsistence agriculture, however, human energy is sometimes the only energy input and it must be counted. If it were not, subsistence farming would show a theoretically infinite efficiency.

In Table 10 some of the results obtained by different researchers are shown. Geographical and climatic differences, as well as differences in the methods of calculation used, prevent the results being strictly comparable. Nevertheless they reveal some interesting points.

For every unit of energy he expends, the Philippines rice farmer obtains a return of 16·5 units and the Tsembaga cultivator of taro-yams in the forests of New Guinea does almost as well. In the UK, on the other hand, in return for a unit of energy expended on

*An average man's output in an eight-hour day of physical work is about a half of a kilowatt-hour – the equivalent of the energy contained in about 4½ tablespoonfuls of petroleum.

Table 10. Efficiency and productivity of some food-producing systems

System	Output kWh/ hectare/ year	Ratio Energy output / Energy input	Output kg protein/ hectare/ year	Input kWh/kg protein
Rice, Philippines*	797·3	16·5	11·4	3·17
Taro-yam gardens, Tsembaga*	388·9	16·4	5·6	4·25
Corn grain, Iowa 1915	7195·0	4·8	167·0	9·06
Corn grain, USA 1945	8945·2	2·02	201·0	22·06
Corn grain, USA 1970	21390·6	2·6	481·0	17·25
Wheat, UK 1970	15612·4	3·4	400·0	11·67
Bread (in shops), UK 1970	12223·2	0·53	350·0	67·51
Main-crop potatoes, UK 1970	15806·8	1·57	376·0	26·67
Pigs in gardens, Tsembaga	861·2	2·1	62·0	11·39
Battery eggs, UK 1970	1944·6	0·14	137·0	98·06
Broiler poultry, UK 1970	1630·6	0·10	203·0	80·56
Milk (average), UK 1970	2778·0	0·37	129·0	57·78
Fishing fleet, UK 1969	—	0·05	—	135·84

Source: G. Leach, *Energy and Food Production*, International Institute for Environment and Development, 1975.

*These use 'slash and burn' methods of cultivation. The land lies fallow for ten to twenty years between crops. Annual yields have been calculated by dividing the yield of a typical crop by 10.

growing wheat the farmer only obtains 3·4 units. And by the time the wheat has been milled, baked, packed and sent to the shop, the energy return per unit of energy input is only 0·53 units.

Turning to intensive agriculture the contrast with subsistence farming is even more remarkable. Battery-egg production consumes seven times as much energy as it yields; battery hens use ten times as much. At the bottom of the list is deep-sea fishing which consumes twenty times as much energy as it produces. The energy productivity of the deep-sea fishing fleet is therefore only 1/330 that of the Philippines rice farmer.

But, it might be argued, energy efficiency is not the full story.

The human diet needs more than energy in its simplest edible form. Protein is necessary too. Looking at the figures for the energy input necessary to produce a kilogram of protein, however, it can be seen that these run roughly parallel to those for energy yield. The Philippines rice farmer is now top of the list. He needs less than a fifth as much energy to produce a kilogram of protein as does the American corn-grower. His methods are 25 times as efficient as those of the UK broiler-poultry farmer. The deep-sea fishing fleet is again the least efficient of all.

Only when the yields per hectare are considered do the advantages of modern farming become apparent. Although the energy-efficiency of American corn-growing has fallen by nearly a half since 1915 the yield per hectare of both edible energy and protein have both roughly trebled. When it comes to using their land the American corn-growers get 27 times as much food energy and 42 times as much protein from the same areas as the subsistence farmer in the Philippines.

High energy inputs to farming are clearly beneficial in that they increase the amount of energy and protein which land can produce. Although they are much less energy-efficient than traditional methods of farming they enable many more people to be supported from the same area of land. They are also very much more productive in labour: a few machines can do the work of an army of labourers. In effect, they turn fossil or other fuels into food. Therein lies the danger.

At this point it is worth turning from how energy is consumed to examine the present state of knowledge about world energy resources and the progress science is making in discovering alternatives to those already known. Soddy observed sardonically: 'It looks as if our successors would witness an interesting race between the progress of science on the one hand and the depletion of natural resources on the other.'[16] The survey of world energy resources in the next part of this book is a look at the progress of that race.

Part Two

Resources

Civilization as it is at present, even on the purely physical side, is not a continuous self-supporting movement . . . It becomes possible only after an age-long accumulation of energy, by the supplementing of income out of capital. Its appetite increases by what it feeds on. It reaps what it has not sown and exhausts, so far, without replenishing. Its raw material is energy and its product is knowledge. The only knowledge which will justify its existence and postpone the day of reckoning is the knowledge that will replenish rather than diminish its limited resources.
– Frederick Soddy, *Matter and Energy*, 1912

Table 11. Geological timescale

Era	Period	Duration millions of years	Began millions of years ago
Cainozoic	Pleistocene	to present	1·0
	Quaternary	1·5	2·5
	Tertiary	62·5	65
Mesozoic	Cretaceous	71	136
	Jurassic	54	190
	Triassic	35	225
Palaeozoic	Permian	55	280
	Upper Carboniferous	45	325
	Lower Carboniferous	20	345
	Devonian	50	395
	Silurian	35	430
	Ordovician	70	500
	Cambrian	70	570
Precambrian			

7

Coal

Coal is a combustible sedimentary rock formed from the remains of plant life. It occurs in recoverable amounts in many areas, though the majority of the world's large coalfields are in the northern hemisphere. It is by far the most plentiful of the earth's fossil fuels.

As it was formed from land-based plants, coal dates back no further than the middle of the Palaeozoic era, about 350 million years ago. The earliest coals belong to the Devonian geological system (see Table 11), but the most prolific period of coal formation was the Carboniferous (345–280 million years ago) when most of the world's hard and bituminous coals were laid down. This period is sometimes called the Anthrocolithicum. Later deposits in the Cretaceous and Tertiary systems formed the lignites and brown coals. Most peat deposits have been formed during the last million years.

In general, the older a coal the greater the extent of change from the original vegetable matter. The process by which the deposit changes to peat, then through brown coal and lignite to increasingly hard bituminous coal and finally to anthracite, is referred to by the inelegant name of 'coalification'. The determining conditions in coalification are temperature and pressure. Since both increase with depth, the high-ranking coals are usually those formed from the older deposits and found at greater depths, although geological upheavals can, of course, bring once deeply buried deposits close to the surface.

In the coal-forming ages the climate was hot and humid and the carbon dioxide level in the atmosphere was probably higher than it is now. These conditions favoured rapid forest growth, and dead vegetation accumulated in the swamps, disappearing under silts and other deposits when land levels sank. Organic matter also accumulated when dead vegetation was carried from the forests by rivers and floods and deposited in lakes and estuaries. This process can still be observed in the deltas of very big river systems like the Mississippi and the Amazon. Over the long stretches of time during which the coal deposits were laid down, there were many major changes in the configuration of the world's lakes, seas and land-masses. In geologically active areas the land might rise and sink many times, allowing only relatively thin seams of vegetable matter to settle, interspersed with a variety of other sediments, whereas in less disrupted areas the vegetable seams were very much thicker.

It is estimated that a 3- to 5-metre depth of peat, which is the first stage in the development of coal, is required to produce a 30-centimetre thickness of coal. The existence of coal seams more than 200 metres thick, although representing massive deposits of vegetation, does not mean that these deposits ever reached thicknesses of up to 1 000 metres at any one time: the lower layers would have been progressively compacted. The rather misleading name of 'coal measures' is sometimes used as a blanket term for all the rock strata laid down in the later Carboniferous ages. References to thicknesses of a thousand metres in the coal measures does not mean that there are coal seams of that size – or indeed any coal at all. The proportion of coal throughout the Carboniferous series of rocks rarely exceeds 2 per cent, so that even in the richest areas coal is a comparatively rare material.

The characteristics of various coals are determined by the types of vegetation making up the original deposits, the kinds of silts washed into the swamps in which decay occurred, the degree to which decay had advanced before the deposit was sealed, and the subsequent geological processes it suffered. The original vegetable deposits consisted principally of carbon, hydrogen and oxygen products produced by photosynthesis. In the process of coalification all these are gradually lost by the elimination of water and carbon dioxide and the evolution of methane (CH_4). Hydrogen and oxygen, however, are lost at a higher rate than carbon, which means that as coalification proceeds the concentration of carbon increases. High-quality anthracite is almost pure carbon. Table 12 shows typical compositions for a number of differently ranked British coals.

Table 12. Analysis of UK coals

	Dry ash-free basis						
	Percentages by weight					Calorific value	
	Carbon	Hydrogen	Oxygen	Nitrogen	Sulphur	Btu/lb	kWh/kg
Wood	49·8	6·2	43·4	0·3	0·3	8400	5·32
Leicestershire (non caking)	78·6	5·2	11·7	1·6	2·8	13990	8·87
Yorkshire (medium caking)	84·8	5·2	7·6	1·7	0·7	15090	9·56
Durham coking coal	88·5	5·0	4·1	1·6	0·8	15630	9·91
South Wales dry steam coal	90·5	3·9	1·4	1·5	0·7	15610	9·89
South Wales anthracite	93·5	3·5	1·0	1·2	0·8	15590	9·88

The properties of different coals define their possible uses. A lot depends, for example, on the amount of sulphur in a coal. A high sulphur content will cause severe pollution problems and it may also preclude the use of such a coal in some processes. Coke, which is used for steel-making, can only be made from a coal which agglutinates or ‘cakes’ satisfactorily when heated. The properties of coking coals are therefore closely defined. Other characteristics of coals which may limit their uses are calorific value, ash content, and the amount of clinker formed in combustion. Near the bottom of the coalification scale are the soft brown coals or lignites. These can contain up to 70 per cent moisture. They disintegrate rapidly in air and are liable to ignite spontaneously when exposed during mining. They are of low calorific value, a half to a third that of good-quality bituminous coal. Their major use is as fuel for power stations. Coal should not be thought of as a uniform fuel capable of being used indiscriminately in any combustion process: rather it is a wide range of different fuels, each suitable for particular uses.

The ease with which any given coal deposit can be mined determines whether or not it can be counted as a resource for human use. There are physical, economic and, ultimately, energetic factors to be considered. In areas which have had a turbulent geological history, the coal seams may be so warped, fractured and faulted, that the coal is, for practical purposes, unrecoverable. Even if the price were right, more energy would be used in digging out rock to get at the coal than could later be recouped in burning that coal.

Coals of high rank are usually deeply buried and have to be mined from underground workings. They are mostly found in seam thicknesses of 3 metres or less – though in exceptional areas the seam thickness may exceed 10 metres. The minimum workable thickness is about 30 centimetres in the most favourable of conditions. Below a depth of about 1 200 metres, high temperatures, increased pressures and geological hazards, as well as the sheer distance of haulage to the surface, rule out the working of any but the richest deposits.

This does not by any means imply that all the coal in a seam of workable thickness at a convenient depth can be recovered. Coal extraction is difficult and a coalmine is a dangerous place. The pressure of the overburden is about 2 tonnes for every metre of depth. Some means of supporting the roof of the workings against such pressure has to be used to allow the coal to be extracted. In the 'long wall' method of mining, which is that mainly used in Europe, the eventual collapse of the roof is accepted. Only the area at the cutting face is supported. As the cutting machines which operate across the whole horizontal width of the coal seam move forward, the roof supports, which are themselves sophisticated, automatically controlled and powered pieces of machinery, also move forward. This leaves the excavated area completely unsupported and the roof slowly falls in, generally resulting in subsidence of the ground above the mine. In the 'room and pillar' method of mining, pillars of coal are left in position to support the roof. The investment in mining machinery and roof supports is less and the method is simpler; it also obviates the problem of ground subsidence. The penalty is that up to half the extractable coal must be left behind.

The list of mining dangers and problems is a long one; under the pressure of the overburden the mine remains in a constant state of movement. The side walls of the tunnels creep inwards and the floor bulges upwards. Newly exposed coal can suddenly release gases which were trapped and compressed within it. Methane, or fire-damp, forms an explosive mixture with air and so can the coal-dust itself. Ground water is another hazard which must be kept under control by constant pumping; and if there are unsuspected weaknesses in the rock, it can burst into the workings with catastrophic effect. In the struggle against the primeval elements – earth, fire, and water – coal is hard won. Mechanization helps but does not eliminate the problems. Machines can only be used where conditions are tolerable for the men controlling them, and they cannot follow coal into the crevices of irregular rock structures.

Open-cast or surface mining is by its nature less hazardous and more efficient in recovery. Peat is extracted by this method and used as a power-station and domestic fuel in Ireland and Russia. In Ireland, where peat is usually referred to as 'turf', the total quantity extracted per year is just under 5 million tonnes – over 30 per cent of the country's electricity is generated in peat-fired power stations. In Russia the quantity used is 57 million tonnes a year. Outside these countries the use of peat for fuel is negligible.

Most open-cast mining is for the lower-ranking lignites and brown coals which are usually found close to the surface and sometimes in very thick layers. One deposit in Victoria, Australia, has a seam over 225 metres thick. In the same coalfield, and in parts of the United States, seams of over 30 metres are found. The scale of operations is often stupendous. In the rich United States fields the overburden may be up to 70 metres thick: this is removed by excavating machines which simply rip up the clays and soft rocks and dump them aside. Explosives take care of harder rocks which are blown up by the acre. Huge draglines, diggers and bucket-wheel excavators shift hundreds of thousands of tonnes of material every day. Large-scale strip-mining is one of the most destructive of man's activities unless rigorous control of operations is maintained and followed by careful rehabilitation of the land. In certain countries, such as the UK, this is now done with some success. But elsewhere, and in the United States in particular, where environmental controls have been adopted belatedly, strip-mining has often left a landscape damaged beyond possibility of repair.

The early importance of coal as a primary fuel for an industrial economy has meant that it has been under detailed study for a long time. The first attempt to produce an estimate of world reserves was made in 1913 at the Twelfth International Geological Conference in Toronto. The tentative conclusion reached was that there was a total of $7{\cdot}397 \times 10^{12}$ tonnes of coal in seams not less than 30 centimetres thick at depths of up to 1 200 metres and in seams not less than 60 centimetres thick at depths between 1 200 and 1 800 metres.

Since then a great deal more has been learned about the earth's geology and its coal resources, though very large uncertainties still remain. Discrepancies, often amounting to many billions of tonnes, are a feature of the various estimates of world coal resources, and even those of individual countries. One reason for this is the cost of detailed exploration. Standard geological surveying techniques will reveal the existence of a coalfield but a great deal more work is required to establish exactly how much coal it contains. As long as a coal-mining country has confirmed figures for reserves sufficient to supply its coal industry for several decades ahead, it has little incentive to make heavy investments in further exploration.

Further confusion arises from the different criteria used to distinguish between 'resources' and 'reserves'. An attempt to produce generally acceptable definitions was made at the World Energy Conference in 1977. 'Geological resources' were defined as coal occurrences which may acquire some economic value for mankind in the future – in other words, it is possible to imagine them being mined some day. Coal reserves, on the other hand, were defined as coal occurrences which are exploitable with present technology and under present economic conditions. As technology improves and the price of competing fuels increases resources will therefore tend to move into the category of reserves.

Tables 13 and 14 provide comprehensive details of geological resources and technically and economically recoverable reserves for all countries presently known to possess exploitable coal deposits. It is noticeable how widespread is the occurrence of coal throughout the world. Intensified exploration in developing countries is now identifying further deposits and there is a cautious optimism that at least some of the world's very poor countries may possess useful amounts of coal.

There is a great difference between the figures for resources and reserves. Present technically and economically recoverable reserves are just about 6 per cent of resources. It will be noticed that the proportion varies considerably between countries. In the US the hard coal reserves are about 10 per cent of resources; in China they are only about 1 per cent. Price, quality of coal, availability of transport and capital equipment, and many other factors determine which coalfields can be mined economically at any particular time.

Although widespread in occurrence, the geographical concentration of coal is also extremely noticeable. The USSR possesses 52 per cent of the world's resources of hard coal and 36 per cent of those of brown coal. The US has nearly 60 per cent of the world's brown coal resources and 15 per cent of the hard coal. The USSR, US and China between them possess 85 per cent of the world's hard coal and 94 per cent of the world's brown coal. This is not to say that the reserves of other countries are negligible. Countries like Germany, the UK, Poland and Australia (to take some of the present largest producers) have resources sufficient to last them hundreds of years at present rates of mining. Some of the developing countries, such as Colombia and India, are capable of meeting their own needs and becoming substantial exporters. Nevertheless, the long-term future of coal production, trade and use in the world is bound to be affected by the decisions made in the three countries with the major part of the world's resources. Only if they are prepared to mine and export quantities of coal far in excess of their own needs can the large-scale expansion of coal use envisaged in some long-term projections of the world's energy future be realized.

As shown in Table 15 the world's present annual production of coal is $3{\cdot}4 \times 10^9$ tonnes (about $2{\cdot}6 \times 10^9$ tce). If this were to remain constant then the world's total geological resources of coal would be exhausted in about 4 000 years. Coal consumption, however, does not remain constant. The table shows how coal consumption has increased since 1932; it has almost quadrupled. Growing populations, increased industrialization and rising living standards all increase consumption of resources. The present growth rate is about 3 per cent per annum. Far from being a dying industry, as is sometimes supposed, coal mining has been growing steadily.

Theoretical studies of possible future patterns of production often extrapolate past growth rates into the future. They show production rising along this extrapolated path until the development of new resources at this rate becomes impossible to sustain. The growth rate then slows and eventually ceases: this is the peak in production. Thereafter output gradually declines. This is called a depletion curve. In his study of world energy resources Hubbert,[17] using a very optimistic figure of $7{\cdot}6 \times 10^3$ billion tonnes of ultimately recoverable reserves, suggests a depletion curve such as that shown in Figure 7. In this, world production grows for the next 175 years, peaks at about eight times the present production level and thereafter declines, with final exhaustion, for practical purposes, occurring in about 800 years. Other depletion patterns could, of course, be considered: higher peaks mean shorter overall durations for the full production cycle; lower peaks allow it to be extended.

When a depletion curve is drawn, the area under the curve represents the total amount of the resource. In Figure 7 it is very noticeable how small a proportion of the total reserves is represented by consumption to date. Humanity has scarcely dented its coal reserves.

This kind of exercise, however, is of limited value. It helps define

Table 13. World coal resources and reserves by coal type – developed and centrally planned countries tonnes coal equivalent × 10^9

	Geological resources				Technically and economically recoverable reserves			
	Hard coal		Brown coal		Hard coal		Brown coal	
Market economies		%		%		%		%
Australia	213·8	2·8	48·4	2·0	18·1	3·7	9·2	6·4
Canada	96·2	1·3	19·1	0·8	8·7	1·8	0·7	0·5
France	2·3	—	—	—	0·4	0·1	—	—
Germany (FR)	230·3	3·0	16·5	0·7	23·9	4·9	10·5	7·3
Japan	8·6	0·1	—	—	1·0	0·2	—	—
Netherlands	2·9	—	—	—	1·4	0·3	—	—
South Africa	57·6	0·8	—	—	26·9	5·5	—	—
Spain	1·8	—	0·5	—	0·3	0·1	0·2	0·1
United Kingdom	163·6	2·1	—	—	45·0	9·1	—	—
USA	1190·0	15·4	1380·4	57·5	113·2	23·0	64·4	44·7
Others	0·7	—	1·7	—	0·2	—	0·6	0·5
TOTAL	1967·7	25·5	1466·7	61·0	239·1	48·7	85·6	59·5
Centrally planned economies								
Bulgaria	—	—	2·6	0·1	—	—	2·2	1·5
China (PR)	1424·7	18·4	13·4	0·6	98·8	20·1	not available	
Czechoslovakia	11·6	0·2	5·9	0·2	2·5	0·5	2·3	1·6
Germany (DR)	—	—	9·2	0·4	0·1	—	7·6	5·3
Hungary	0·7	—	2·8	0·1	0·2	—	0·7	0·5
Korea (DPR)	2·0	—	—	—	0·3	0·1	0·2	0·1
Poland	121·0	1·6	4·5	0·2	20·0	4·1	1·0	0·7
Romania	0·6	—	1·3	—	—	—	0·4	0·3
USSR	3993·0	51·7	867·0	36·1	82·9	16·8	27·0	18·8
Others	0·2	—	—	—	0·1	—	—	—
TOTAL	5553·8	71·9	906·7	37·7	204·9	41·6	41·4	28·8
TOTAL DEVELOPED AND CENTRALLY PLANNED	7521·5	97·4	2373·4	98·7	444·0	90·3	127·0	88·3

Table 14. World coal resources and reserves by coal type – developing countries tonnes coal equivalent × 10^9

	Geological resources				Technically and economically recoverable reserves			
	Hard coal		Brown coal		Hard coal		Brown coal	
		%		%		%		%
Botswana	100·0	1·3	—	—	3·5	0·7	—	—
Mozambique	0·4	—	—	—	—	—	—	—
Nigeria	—	—	0·2	—	—	—	—	—
Zimbabwe	7·1	0·1	—	—	0·8	0·2	—	—
Swaziland	5·0	0·1	—	—	1·8	0·4	—	—
Zambia	0·2	—	—	—	—	—	—	—
Others	2·4	—	—	—	1·0	0·2	—	—
TOTAL AFRICA	115·1	1·5	0·2	—	7·1	1·5	—	—
Bangladesh	1·6	—	—	—	0·5	0·1	—	—
India	55·6	0·7	1·2	0·1	33·3	6·8	0·3	0·3
Indonesia	0·6	—	3·1	0·1	—	—	1·4	0·9
Iran	0·4	—	—	—	0·2	—	—	—
Korea (R)	0·9	—	—	—	0·4	0·1	—	—
Turkey	1·3	—	2·0	0·1	0·1	—	0·7	0·5
Others	5·4	0·1	0·4	—	1·5	0·3	—	—
TOTAL ASIA	65·8	0·8	6·7	0·3	36·1	7·3	2·4	1·7
Argentina	—	—	0·4	—	—	—	0·3	0·2
Brazil	4·0	0·1	6·0	0·3	2·5	0·5	5·6	3·9
Chile	2·4	—	2·1	0·1	—	—	0·1	0·1
Colombia	7·6	0·1	0·7	—	0·4	0·1	—	—
Mexico	5·4	0·1	—	—	0·9	0·2	—	—
Peru	1·0	—	—	—	0·1	—	—	—
Venezuela	1·6	—	—	—	1·0	0·2	—	—
Others	—	—	—	—	—	—	—	—
TOTAL LATIN AMERICA	22·3	0·3	9·2	0·4	4·9	1·0	6·0	4·2
Yugoslavia	0·1	—	10·8	0·4	—	—	8·4	5·9
TOTAL DEVELOPING COUNTRIES	203·3	2·6	27·0	1·1	48·2	9·8	17·0	11·8
WORLD TOTAL	7724·8	100	2400·4	100	492·5	100	143·9	100

Table 15. Growth of world production of coal and lignite – 1932–77

Date	Coal	tonnes × 10⁶ Lignite	Total
1932	870	164	1 034
1937	1 154	233	1 387
1942	1 291	314	1 605
1947	1 204	237	1 441
1952	1 500	435	1 935
1957	1 735	593	2 328
1962	1 857	684	2 541
1967	1 949	722	2 671
1973	2 164	881	3 045
1977	2 475	946	3 421

Source: *UN Statistical Year Books.*

outer limits and it provides a context for discussion. But it certainly does not indicate what must take place, or even what is possible, or likely. It enables it to be said, for instance, that a world economy based on expansion of coal production at rates extrapolated from the past cannot be sustained for much more than the next century and a half. But it does not justify anyone going much beyond this kind of generalized statement. The world is not a single, uniform entity as far as the distribution of coal reserves is concerned. To aggregate all reserves and treat them as though they were a uniform-

---

Source for Tables 13 and 14: *Coal Development Potential and Prospects in the Developing Countries*, World Bank, October 1979.

Note: Since the quality of coal varies so much, attempts have also been made to bring some uniformity into national and international statistics. The World Energy Conference divides coal into two categories on the basis of its heat content. Hard coal includes anthracite and bituminous coal with a calorific value greater than 5 700 kilocalories per kilogram; brown coal includes sub-bituminous coals and lignites with a heat content lower than this figure. Within these categories actual tonnages of coal are converted to tonnes of coal equivalent (tce) on the basis of their individual heat contents – the standard tonne of coal equivalent has a heat content of 7 000 kilocalories per kilogram.

Figure 7. Theoretical depletion curve for $7{\cdot}6 \times 10^3$ billion tonnes recoverable reserves of coal

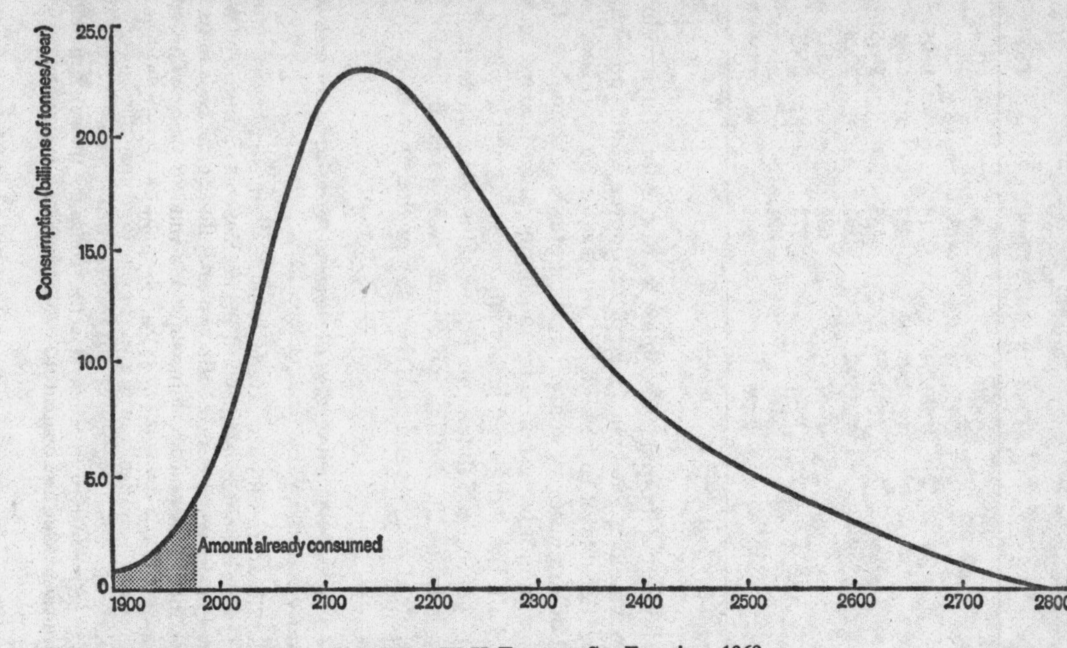

After M. K. Hubbert in *Resources and Man*, W. H. Freeman, San Francisco, 1969.

ly distributed 'world resource' equally available to anyone wishing to consume them is quite unrealistic. It is a serious flaw in many 'futures' studies, both optimistic and pessimistic, that they aggregate world totals of various resources and treat them in this way.

The future use of coal cannot be predicted by any such simple model. Estimates of coal reserves are accurate only within a range of thousands of millions of tonnes – even within the margin of error the quantities are sufficient to supply the world for hundreds of years at its present rate of consumption. There is so much coal that speculation about its ultimate exhaustion is irrelevant. But its distribution is very uneven, and this will profoundly affect its future use. There is no average or typical country. The whole southern hemisphere has only three-quarters the resources of Europe; southern and central America have only an eighth of the resources of Australia. And the picture is completely dominated by three countries: the USSR, the USA and China.

Table 16 shows the present pattern of world coal production. Future changes will be determined by the complex interaction of a large number of factors. The traditional coal-mining countries of western Europe have been running down their coal industries in recent decades. In the UK, for instance, production has almost halved, falling from 214 million tonnes a year in 1954 to 120 million tonnes a year in 1978 – during the same time the number of miners fell from 700 000 to 233 000. In France production fell from 73 million tonnes in 1952 to 25 million tonnes in 1977; and in Holland, where production was over 12 million tonnes in 1950, coal mining has now ceased completely. This kind of decline is difficult to arrest, let alone reverse. Work-forces are hard to reassemble once the traditional mining communities have been dispersed. Mines, once closed down, suffer collapses and flooding and are usually impossible to reopen. Opening new pits and creating new distribution and storage facilities in crowded, highly urbanized countries is almost prohibitively difficult. Coal mining brings slag heaps, polluted waterways, dirt, heavy road and rail traffic and a big loss of otherwise usable land. In urban areas there is also the problem

Table 16. World production of hard and brown coal – major producers, 1977
Tonnes × 10^6

Country	Hard coal	Brown coal	Total	%
USA	608	15	623	22·5
USSR	419	97	516	18·6
China (PR)	500	—	500	18·0
Poland	186	11	197	7·1
UK	123	—	123	4·5
Germany (FR)	84	37	121	4·4
India	99	1	100	3·6
Czechoslovakia	28	56	84	3·0
Australia	71	11	82	3·0
South Africa	81	—	81	2·9
Germany (DR)	—	76	76	2·8
Korea (DPR)	43	7	50	1·8
Canada	23	3	26	0·9
France	22	2	24	0·9
Yugoslavia	1	19	20	0·7
Japan	18	—	18	0·7
Korea (R)	17	—	17	0·6

Source: *Coal Development Potential and Prospects in the Developing Countries*, World Bank, October 1979.

of ground subsidence after the coal has been extracted. In some cases the ground may sink as much as a metre. This plays havoc with underground services such as sewers and water pipes. It also necessitates the replacement of damaged buildings. All these features are part of the way of life in a mining community, which society generally manages to ignore. It is difficult to see them being easily accepted in areas which have not previously experienced them. Evidently, there are formidable pressures against any major expansion in the European coal-mining industry.

In other parts of the world there are greater possibilities for expanding production. The World Coal Study, which published its findings in 1980,[18] envisaged world coal production at least doubled, and possibly trebled, by the year 2000. These figures are based on an assessment of the rate of increase in world energy demand and the role coal might play in meeting it. The hypothesis is that coal will take over from oil as the world's main source of supply for whatever growth in energy consumption occurs over the next few decades.

At the moment most coal is consumed in the country of its production. Expanding use on the scale described in the World Coal Study would require a large growth in international coal trade. At present this is about 200 million tonnes a year, of which only 50 million tonnes is steam coal (for electric power generation and industrial boilers and furnaces); the rest is metallurgical coal, that special type of coal used to manufacture coke for steel-making. The use of coal in the industrialized countries on the scale envisaged would require that steam-coal trade multiply by ten to fifteen times.

Under such conditions and assuming, as now seems likely, that neither China nor the USSR become large exporters to the Western world the major exporting countries would be Australia, Canada, South Africa and the USA. Of these by far the most important would be the USA. This would restore the USA to the position in the world energy market it previously held as the dominant force in the international oil trade. It is noticeable that the major oil companies are already acquiring large holdings of coal reserves both in the USA and abroad. Exxon is investing, for example, in Colombia; Shell is promoting developments in Botswana and Swaziland. A new investment force, however, is Japan with a programme to subsidize coal research projects abroad and to guarantee loans for coal development projects in foreign countries. By these means the Japanese government hopes to secure its country's long-term coal supplies.

Obviously the possibility of substantial expansion in world coal production needs to be taken seriously. There is much to be said for it as a means of lessening dependence on oil. But the environmental record of coal has been a sorry one of smoke-blackened towns, devastated landscapes, polluted rivers, and mineworkers with broken health. Most of these problems can be dealt with satisfactorily – at a price. Landscapes can be restored after mining. Under proper working conditions mining need be no more dangerous than any other heavy industry.

The pollution problem is the most intractable. It is also very complex, for depending on the content of the coal and the method of burning, the following pollutants may be emitted: sulphur oxides, heavy metals (lead, mercury, cadmium), radioactive elements such as radium and uranium, organic carcinogenic compounds such as polycyclic hydrocarbons and, of course, plenty of carbon dioxide.

Although the installation of pollution control devices can control the emission of, say, sulphur oxides, the device itself sometimes ends up causing its own environmental problem. When sulphur is removed from power-station flue gases by an extractor or 'scrubber' using limestone to absorb the sulphur, then hundreds of thousands of tons of very noxious limestone-sulphur sludge must somehow be disposed of. The quantities of heavy metals, radioactive and carcinogenic substances emitted are still small and their effect on the environment is not fully understood, although it may well be cumulative. These emissions would clearly be considered a potential menace if coal use were greatly increased.

Carbon dioxide is emitted in the respiration of living creatures, and all hydrocarbons, wood, petroleum and coal, produce carbon dioxide on burning. But because coal is mainly carbon, it releases 25 per cent more carbon dioxide than oil and 75 per cent more than natural gas per unit of heat produced. It is now generally agreed that the carbon dioxide content of the atmosphere has been increasing at about 0·4 per cent per year and that, if this were to continue, it could have a warming effect on the world's climate. The quantities released by coal combustion are as yet relatively small in relation to the total quantities of carbon dioxide in the atmosphere, in solution in the oceans, and in the biological systems of the earth. The World Climate Conference at its meeting in Geneva in February 1979 said of the growth in the carbon dioxide concentration in the atmosphere:

It is possible that some effects on a regional and global scale may be detectable before the end of this century and become significant before the middle of the next century. This time scale is similar to that required to redirect, if necessary, the operation of many aspects of the world economy, including agriculture and the production of energy.

While there is no apparent need to restrict coal use over the next few decades, continuous monitoring and further research is obviously very necessary. The global carbon cycle has yet to be accurately and minutely described. In the longer term, the control of carbon dioxide in the atmosphere may turn out to be one of the most grievous problems facing the human race. All in all, coal-fired power stations are undoubtedly more polluting, and may even be more radioactive, than a smoothly running nuclear power station. This fact has been noted, with no little satisfaction, by proponents of nuclear power.

At present coal supplies about 30 per cent of the world's energy requirements. It can provide a substitute for oil in many uses in the developed countries; in some developing countries it may become the only fuel available for electric power generation and industrial development. This would call for enormous investment in mines, transport systems and environmental research, quite apart from the cost of even partially adapting to coal in economies which have evolved round the use of oil. An increase in coal use is now being considered with great seriousness by energy policy-makers everywhere. Its magnitude will be determined by the general growth in energy demand, and the competitive position of coal in relation to other available energy sources.

8

Petroleum

The word petroleum, meaning 'rock oil', is now used as a blanket term for a wide range of hydrocarbons – from the simple gas methane, or natural gas, through liquids of increasing viscosity to the solid paraffin waxes. Petroleum has been used as a fuel, lubricant, medicament and even building material for thousands of years. The biblical 'burning fiery furnace' was probably a lighted natural-gas escape, as were the eternal flames at various oracular shrines. In the Middle East bitumen has traditionally been used as a mortar, flooring and waterproofing material – it was reputedly used in the construction of the ill-fated Tower of Babel. It was also used in the preservation of mummies. Oil has also been used, more or less harmfully, as a salve or internal medicament for all kinds of ailments. It is now one of the basic raw materials of the pharmaceutical industry.

The origin of petroleum has long been a subject of speculation. Under high temperatures carbon enters into combination with various metals to form carbides which are decomposed by water to give hydrocarbons such as acetylene (C_2H_2); it therefore once seemed reasonable to suppose that petroleum products were formed in this way deep in the earth's crust. A case has even been made for the extra-terrestrial origin of at least some of the earth's hydrocarbons.[19] The tails of comets are composed of solid methane and the gas is a constituent of the atmosphere of some of the other planets. However, there is now little doubt that the great majority of the earth's petroleum is derived from marine organic material, laid down as part of the accumulation process by which the world's sedimentary rock basins were formed. The earliest oils, found in the lower Palaeozoic beds, or even in some of the Precambrian formations, predate the formation of coal by hundreds of millions of years and were laid down long before the appearance of land vegetation.

Although the formation sequences of coal and oil have some common features there are equally important differences and the two materials are always stratigraphically separated. The precise development of each one of the multitude of different petroleum substances found in any one deposit is almost impossible to trace through the process in which the various marine organic deposits accumulated, underwent partial bacterial decay, and were sealed by subsequent inorganic deposits. There is evidence that the initial material formed may have been kerogen, a hard dark waxlike solid found, for instance, in oil shales, and that this was broken down by subsequent heat and pressure. Temperature is a critical factor, since many liquid-petroleum substances break down at about 150°C. Because of the temperature rise associated with depth, very little liquid petroleum is found below 7 600 metres.

Whereas coal is found in the geological stratum in which it was formed, petroleum usually migrates from the place of its origin. Under pressure from the overburden, liquid and gaseous hydrocarbons are displaced from the rocks in which they were formed and diffuse through the surrounding permeable strata until they are trapped, still under pressure, by an impermeable barrier. If no trap is encountered the petroleum escapes to the surface where it is eventually broken down into its original constituents. The development of an oilfield thus has to be distinguished from the formation of the petroleum itself. The liquids and gases found in an oilfield do not necessarily have a common origin; the methane may, for instance, have come from the formation of coal. What they have in common is that their migratory progress through various permeable rock formations has ended in the same place.

The formation of an oilfield in any area therefore depends on the existence of suitable geological traps to which the petroleum can migrate and in which it can accumulate. But later faulting, folding, or igneous intrusions can destroy or dissipate these accumulations. Such is the rarity of the coincidence of all the necessary conditions that only about one in twenty of the structures identified from geological evidence as being suitable as an oil reservoir actually yields petroleum in recoverable amounts.

Oil

The main blame for one of the commonest misconceptions about oilfields can probably be laid on Jules Verne's *Journey to the Centre of the Earth*. The great underground caverns he so vividly describes seem ideal for the accumulation of lakes of oil. In fact, oil accumulates within the pore spaces of a variety of sedimentary rocks, such as the sandstones, or carbonates like limestones or dolomite. The volume of pore space available to be filled varies from about 35 per cent down to as little as 5 per cent of the total rock volume. Extracting oil is more like squeezing treacle out of a brick than lifting bucketsful of water from a well.

Within a reservoir, the oil, gas and water tend to separate into layers with gas at the top, oil next, and then water. The extent of this separation depends on the porosity of the rock, the viscosity of the liquid petroleum and the relative proportions of the liquids and gases. When the cap over the reservoir is punctured by a drill, the pressure is released and the liquids and gases shoot to the surface. In early years of oil exploration this frequently produced a 'gusher' which was completely uncontrollable until the pressure dropped. In these cases most of the natural pressure of the well was lost and the amount of oil obtained was much less than it might have been. Obtaining the maximum yield from any oilfield requires careful management of the pressure by the spacing of wells and the control of rates of extraction. It is possible to mismanage the process so badly that the reservoir pressure is dissipated in driving up briny water, leaving the greater part of the oil deposit irrevocably trapped.

The extraction of oil using only the natural reservoir pressure has the technical name of 'primary recovery'. Wells used to be abandoned when the natural flow ceased, but it was discovered that injecting gas or water under pressure could restore the flow and this came to be called 'secondary recovery'. Efficient modern practice does not, however, allow pressures to decline to zero before commencing 'secondary' methods; the objective is to maintain the pressure at an optimum level for as long as possible. Often natural gas is reinjected into the reservoir from which it has been obtained in order to maintain the pressure for oil extraction. Gas injection was first used in the United States as early as 1891 but has only come into widespread use since the end of the Second World War. Secondary recovery by water flooding is also used, but it is more difficult to engineer and is not possible in many areas because of the amount of water required.

More elaborate 'tertiary' recovery methods use steam, solvents or underground combustion, fed with air or oxygen from the surface, to reduce the viscosity of the oil in the reservoir and allow it to flow more freely. The use of heat from underground nuclear explosions has also been suggested, but no way of preventing the oil from becoming radioactive, and hence unusable, has yet been convincingly demonstrated.

Viscous oils in rocks of low porosity have very low recovery factors, as little as 5 per cent. Even under ideal conditions recovery factors are rarely more than about 70 per cent. The average recovery factor in the US is about 33 per cent, in the North Sea about 45 per cent.

Discussions about the exhaustion of the world's oil can cause considerable confusion. Some of it would be avoided if it were always clear what people were talking about. Here is one quotation:

> If these rates continue to grow exponentially, as they have done since 1960, then natural gas will be exhausted within fourteen years and petroleum within 20 years.

And a contrasting view that

... the oil resource base in relation to reasonable expectations of demand gives very little apparent cause for concern, not only for the remainder of this century but also thereafter well into the 21st century at rates of consumption which will then be five or more times their present level.

The first is from 'A Blueprint for Survival',[20] not perhaps the most optimistic document of recent times, and the second is from a paper[21] by Professor Peter Odell who may well be the most optimistic oil economist of all time. Quite apart from their undoubted ideological differences and penchant for looking only at the dark or bright sides of the same cloud, they were actually talking about two very different things. 'A Blueprint for Survival' was referring to published figures for 'proved reserves'; Professor Odell was talking about the 'resource base'.

The 'resource base' is defined as all the petroleum within the earth's crust in a particular area. It is sometimes called the 'oil in place'. Since recovery rates vary from 5 per cent up to 70 per cent of the oil in a reservoir, 'resource base' figures must always be qualified by reference to an expected rate of recovery.

On the other hand 'proved reserves' or, as they are sometimes called, 'published proved reserves' are defined by the oil industry as 'the volume of crude oil which geological and engineering information indicate beyond all reasonable doubt to be recoverable from an oil reservoir under existing operating conditions'. This is a very restricted definition. Its main purpose is to prevent fraudulent raising of money by oil companies publishing over-optimistic figures for their assets in the ground. There is nothing hypothetical about a proved reserves figure. It is, in effect, the verified inventory of an oil company's stock in hand.

The proved-reserves figure understates the true recoverable reserves in a number of ways. It is confined to oil which has been identified as recoverable in an oilfield: it does not include oil which may lie in extensions of the field yet to be explored nor does it include oil which may be recoverable using secondary and tertiary methods until these are actually in operation. The figure for proved reserves is therefore one of limited applicability. At the beginning of the exploration of an oilfield it will necessarily understate the amount of recoverable oil. As exploration continues, an increasing proportion of the recoverable reserves will be identified until, finally, the estimate for remaining proved reserves will coincide with the figure for ultimately recoverable oil. A study published by the Alberta Energy Resources Conservation Board has shown, from the analysis of the production records of 128 oilfields in the province, that the quantity of oil finally recovered from each field was, on average, about nine times the published reserves declared at the end of the first year of exploration. The Alberta data can hardly be applied directly to other areas, but the general principle of the appreciation of the initially published reserves figure is certainly valid.

Table 17 shows the present distribution of the world's proved reserves. As in the case of coal, the unevenness is very marked. The Middle East has 56 per cent of the total and a further 9 per cent are in Africa. The United States, which is the world's third largest producer, has only 4 per cent. The UK sector of the North Sea has over 2 per cent, a small proportion of world reserves, but over half the US total. Over the past five years the world's total proved remaining reserves have fallen by more than 10 per cent. New discoveries are failing to keep pace with the rate of depletion.

Much work has been done on the estimation of total ultimately recoverable reserves. Two approaches are used: the geological and the statistical. The first depends on the present sophistication of geological maps of the world. The main sedimentary basins have all been identified by now and most of them have been subjected to at least some exploratory drilling. Experience of developed oilfields can therefore be used as a guide to the prospects in areas yet to be explored in detail. Some will be more prolific than expected, as in the case of the North Sea; others, such as Alaska, probably less so. In an unbiased calculation such errors should tend to cancel out. With time, as drilling activity reduces the uncertainties, this method should provide estimates which increasingly converge on the truth.

Table 17. World proved oil reserves – selected countries, 1979

Country/region	Reserves barrels $\times 10^6$	% of world total
ASIA–PACIFIC		
Australia	2 130	0·33
Brunei	1 800	0·28
India	2 600	0·41
Indonesia	9 600	1·50
Malaysia	2 800	0·44
TOTAL: ASIA-PACIFIC	19 355	3·02
WEST EUROPE		
Norway	5 750	0·90
United Kingdom	15 400	2·40
TOTAL: WEST EUROPE	23 776	3·71
MIDDLE EAST		
Abu Dhabi	28 000	4·36
Dubai	1 400	0·22
Iran	58 000	9·04
Iraq	31 000	4·83
Kuwait	65 400	10·19
Neutral Zone	2 260	0·35
Oman	2 400	0·37
Qatar	3 760	0·59
Saudi Arabia	163 350	25·46
Syria	2 000	0·32
TOTAL: MIDDLE EAST	361 947	56·41
AFRICA		
Algeria	8 440	1·32
Angola-Cabinda	1 200	0·19
Egypt	3 100	0·48
Libya	23 500	3·66
Nigeria	17 400	2·71
Tunisia	2 250	0·35
TOTAL: AFRICA	57 072	8·89
WESTERN HEMISPHERE		
Argentina	2 400	0·37
Brazil	1 220	0·19
Ecuador	1 100	0·17
Mexico	31 250	4·87
Venezuela	17 870	2·79
United States	26 500	4·13
Canada	6 800	1·05
TOTAL: WESTERN HEMISPHERE	89 773	13·99
COMMUNIST AREAS		
USSR	67 000	10·44
China	20 000	3·12
TOTAL: COMMUNIST AREAS	90 000	14·03
TOTAL: WORLD	641 624 (88×10^9 tonnes)	

Note: Only countries with reserves over $1\,000 \times 10^6$ barrels are listed in the above table.

Source: *Oil and Gas Journal*, 31 December 1979.

The statistical approach is based on an analysis of the relationship between the amount of exploratory drilling and the rate of discovery of oil. In any oil-bearing area it would be expected that the large, easily discovered fields would be found first. Subsequent discoveries then become more difficult and tend to require a greater amount of drilling. As the oilfield is explored the rate of discovery per given amount of drilling will tend to fall. In theory, therefore, it might be possible to predict from the drilling and discovery rates how much oil remains to be found. Analysis of areas which have already been explored heavily has tended to confirm this and some mathematical relationships have been derived. An astonishing amount of data is available in the United States which has had over a century of exploration and at present has no less than 508 000 operating oilwells, out of a total of 582 000 in the whole of the non-Communist world.

The most interesting use of this technique has been made by M. K. Hubbert in the United States. In 1956 he predicted that United States oil production would reach its peak during the period 1966–71. He later amended this slightly and said peak production of about 410 million tonnes a year 'should be expected to occur about 1969 plus or minus a year or two'.[17]

The authors of a study of methods of estimating reserves which was published as late as 1965 commented that they

> ... find it impossible to take seriously an estimate of the future availability of crude oil based solely on a mathematical formula for projecting the statistical history of cumulative additions to gross reserves . . . Mr Hubbert's work with numbers and techniques appears to add nothing to the embryonic science of petrolumetrics.

They did, however, commend a report by C. L. Moore: 'Mr Moore utilizes all the data available and his study must be regarded as a useful attempt to advance knowledge by employing trend analysis.'[22]

Moore forecast a peak in US production in 1990 at a production level of about 680 million tonnes a year. In fact, US crude-oil production reached a peak of 478·6 million tonnes in 1970 and has slowly declined since then. Whether or not the criticisms of his methodology are justified, Hubbert has the considerable moral advantage over his critics of being proved right, thus far, by events.

Table 18 shows various estimates of ultimately recoverable oil reserves for the world which have been made since the 1940s. The spread of these is large, with a factor of four between the highest and the lowest. But an important point should be noted: the unique magnitude of the Middle East fields was not fully appreciated until

Table 18. Estimates of world ultimately recoverable crude-oil reserves

Date	Source	Estimate barrels × 10⁹
1946	Duce	400
1946	Pogue	555
1948	Weeks	610
1949	Levorsen	1 500
1949	Weeks	1 010
1953	MacNaughton	1 000
1956	Hubbert	1 250
1958	Weeks	1 500
1959	Weeks	2 000
1965	Hendricks	2 480
1967	Ryman	2 090
1968	Weeks	2 200
1969	Hubbert	1 350–2 100
1970	Moody	1 800
1971	Warman	1 200–2 000
1971	Weeks	2 290
1972	Warman	1 900
1972	Bauquis, Brasseur and Masseron	1 950
1975	Moody and Geiger	2 000
1978	Nehring	2 025

the middle 1950s. Until then most of the detailed information on oil-bearing rock formations had come from borehole records available in the United States. These were used to make estimates of the likely productivity of oil-bearing sedimentary basins in other parts of the world. Researchers failed to foresee how radically the Middle East was going to break the established rules. Later estimates incorporate the Middle East data. Looking at the estimates which have been made over the past ten years the spread is much narrower, with a clustering of the more recent figures around a total of about 2 000 billion barrels.* The various researchers,

*Of this total of 2 000 billion barrels, or 274 billion tonnes, 55 billion tonnes have already been consumed; 88 billion tonnes are published proved reserves; and 131 billion tonnes remain to be discovered.

using a variety of approaches, have produced results which closely coincide with each other. The margin of error in any of these estimates is, of course, high. But the convergence of results around 2 000 billion barrels lends support to the view that this may be near the true figure. If so, then ultimately recoverable oil reserves are about one twenty-fifth those of coal in terms of heat equivalent.

Such estimates have their critics. One of the most vociferous is Professor Peter Odell who would not concede that the figure of 2 000 billion barrels is anywhere near the truth. He believes, with considerable justification, that oil companies deliberately conceal information about their reserves from governments and the public. Odell has suggested that it would be legitimate to treble this figure;[21] but his arguments have been heavily criticized.

A prudent course at the moment is to assume that the generally accepted figure of 2 000 billion barrels is the more likely to be right. It is worth remembering, too, that of this around half remains to be discovered. If present production levels are to be maintained, this is the equivalent of finding new oilfields equivalent to the North Sea every two years for the next fifty years. Put that way, the task is formidable indeed and some knowledgeable oil geologists, reviewing the poor exploration results of the past decade, believe it to be beyond the power of the world's oil industry.

The question is sometimes asked whether the huge rises in oil prices do not introduce a new element into oil exploration and production. Might not the oil companies, stimulated by higher prices, search for and produce oil which previously would not have been classed as recoverable? This argument is passionately supported by the oil companies who frequently complain about their 'lack of incentive' to invest in drilling and producing when prices are low or taxes high. Unfortunately, there is little substance to this apparently strong case. The awesome difficulties of the North Sea and Alaska's Prudhoe Bay were tackled when oil prices were just a tenth of today's. Large oilfields are attractive to oil companies at virtually any price; increasing the price of oil does not, however, make them any easier to find.

Since the beginning of oil exploration there have been about 30 000 'significant' oil discoveries. But of these just thirty-three 'supergiant' fields – that is, fields containing more than 5 billion barrels – contain about a half of all the oil discovered; a further 250 'giant' fields – that is, fields with more than 500 million barrels – contain a further 25 per cent of the world's oil. The remaining 25 per cent is divided between the other 29 700 discoveries.

Increases in the price of oil therefore only have an effect on decisions about whether or not to produce from fields at present deemed uncommercial. Although such decisions may have considerable importance to an oil company or a small country, the yield from small fields has little bearing on the world's total oil production.

But higher prices could increase the length of time a field is kept in production. A typical large oilfield is brought up to full production over a period of four or five years, and held at this for another four or five years, after which production declines at perhaps 10 per cent per annum. Theoretically, there is always some oil left to extract. For the Californian with a stripper well in his garden and negligible production costs it is worth keeping it going as long as there are a few barrels to take round to the local refinery every couple of weeks; higher prices would merely make this more rewarding. Maintaining a manned platform in the North Sea, however, demands a much higher rate of financial return. In this case, higher prices would undoubtedly make it possible to continue production for a longer period. But being the tail-end of production, this would add only a small proportion to the total oil recovered from such fields, certainly not enough to make a major difference to the world's total recoverable reserves.

Figure 8 shows a theoretical depletion curve for world oil reserves of 274 billion tonnes (2 000 billion barrels). This kind of curve was common in discussions about the future of oil some years ago. It shows a peak in world oil production in the 1990s and a gradual decline after that as the world accustoms itself to doing without oil. Up to 1973, world oil consumption had grown at an

Figure 8. Theoretical depletion curve for world oil reserves of 274 billion tonnes

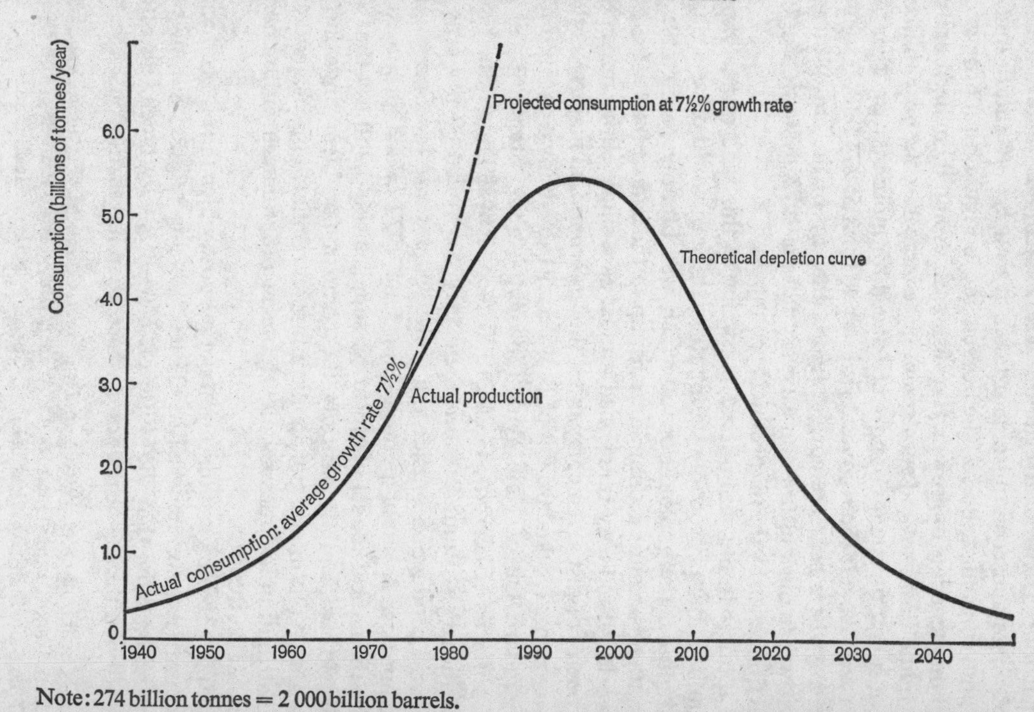

Note: 274 billion tonnes = 2 000 billion barrels.

uncannily uniform rate of about $7\frac{1}{2}$ per cent per annum – production had doubled with such regularity every eight or nine years that many planners and economists had taken it as a law of nature that it could continue to do so. The broken line in Figure 8 is a projection of the same rate of growth. It reveals that, even if there had been no Middle East war and subsequent price increases, the growth in oil consumption could not have continued for more than another five or ten years. If the recoverable reserves' total of 274 billion tonnes is anywhere near the correct figure, the end of the era of steady growth in oil consumption was very close in 1973, even without the action taken by the oil-producing countries.

But, as in the case of coal, the value of this kind of analysis is limited. It was probably legitimate to think of world oil as a single resource, the exploitation of which was broadly predictable in mathematical or economic terms, up to the end of the 1960s. The world oil industry functioned almost as a single entity under the control of just seven companies, the 'international majors' as they were called: Standard Oil of New Jersey (now known as 'Exxon'), Standard Oil of New York (Socony, Mobil), Standard Oil of California (Chevron), Gulf Oil, Texaco, Royal Dutch Shell and British Petroleum. Between them they controlled 80 per cent of world production outside the United States and the Communist countries. A study published in June 1973 which proposed a mathematical model of world oil supplies could still say that the world problem 'is basically the company problem on a larger scale'.[23] The international oil companies regulated prices, markets, distribution facilities and development programmes with remarkable effectiveness.

The cohesion and uniformity of purpose which the oil-producing countries have achieved in OPEC* during recent years means, paradoxically, that the world oil market is very much less of an entity, in the classic economic sense, than it was previously. In the

*Organization of Petroleum Exporting Countries: Ecuador, Venezuela, Iran, Iraq, Kuwait, Qatar, Saudi Arabia, United Arab Emirates (Abu Dhabi, Dubai and Sharjah), Algeria, Libya, Gabon, Nigeria and Indonesia.

study mentioned above the authors said: 'The main hypothesis offered for consideration is one that has long been familiar in classical economics. The hypothesis is that, at equilibrium, prices are equal to marginal costs.' This is no longer valid. An oil-producing country with control over its own resources will produce as much oil as it requires for its own internal consumption. On top of that it will produce for export as much as is necessary to pay for its imports and build up the foreign currency reserves it feels it needs. Only by coincidence will these quantities add up to the amount predicted by the theoretical economic analysis. Price rises enable the same results to be achieved with lower levels of production and hence slower depletion of reserves. This may be contrary to the classic theory which assumes that higher prices encourage higher rates of production, but it is thoroughly in accordance with common sense. The oil countries are well aware of the absurdity of embarking on programmes of industrialization while at the same time rapidly depleting the resources on which these industries will depend.

It is now the declared policy of most of the OPEC producers not to increase their production any further. The world has thus probably reached its peak in oil consumption; there is little ground for supposing that world oil production will ever rise much beyond its present level of 3·2 billion tonnes a year. More importantly, the amount of oil offered for international trade is likely to decrease since the OPEC countries are gradually building up their own consumption as their economies develop. At the moment, international trade in oil is about 35 million barrels per day (1 700 tonnes per year), of which about 28 million barrels come from the OPEC countries.

From the early 1960s the increased energy requirements for economic growth and industrial development have mostly been met by oil. In the industrial countries of the Organization for Economic Cooperation and Development (OECD) oil provided nearly 70 per cent of the growth in energy consumption between 1960 and 1978. In the developing countries the proportion was generally even higher. Moreover oil was the most flexible of the world's energy sources: when energy demand slackened or picked up the oil taps could be turned down or up to match. All this changed in 1973, but the combined effects of the economic recession which slowed demand and the substantial contributions of the Alaska and North Sea oilfields masked the full significance of what had happened. The period of 1974 to 1978 in which oil prices, in real terms, actually went slightly down was a false summer indeed.

The Iranian revolution in late 1978 revealed the true underlying instability of the world oil market. Although as a proportion of total world production the loss of Iranian oil was small, about 5 per cent, it jolted the market into a state of turbulence. Acute disagreements emerged in OPEC. They did not, however, presage its disintegration and the collapse of the oil price, as some devotees of the classical theory of cartels had foretold. Quite the contrary: the prices charged by the various producers all moved upwards, at different rates but always rapidly. By the beginning of 1980 the $12–13 a barrel which oil had been fetching just a year earlier had risen to $30–35 and showed no signs of stopping there for any prolonged period.

Market forces have not been entirely banished from the oil market, but the conditions under which they will operate have changed. The prospect therefore is one of oil prices rising at a pace sufficient to prevent any further increase in demand, or rising until economic recession, itself perhaps caused by oil prices, reduces that demand. Within this context, oil will tend to go to those countries best able to pay for it, and to those uses in which it can command the highest prices. The prospect for the oil importing countries, whether industrialized or developing, is therefore one of increasing competition for the declining total of available oil. Any country which increases its share will do so only by decreasing the amount the others can have. It is a harsh outlook, fraught with possibilities for international tensions and social divisions.

The energy 'crisis' has been widely misconstrued as implying that the world will run out of oil within the next couple of decades.

The reality is less dramatic but very serious, for there is no satisfactory substitute for oil – certainly none that can be transported so safely, that flows conveniently in pipes, and suits existing equipment. The assumptions on which were based so many of the plans and hopes of people all over the world for a continuation of the industrial progress of the past few decades are in need of drastic revision.

Natural Gas

Crude oil in a reservoir is always accompanied by a mixture of gases which, for convenience, is termed 'natural gas'. Its composition varies but the principal component is methane (CH_4), which usually forms between 85 and 95 per cent of the total. The remainder is mainly composed of higher hydrocarbons such as ethane, propane and butane. Helium, nitrogen and some of the rare gases like argon are also found. One of the most important non-hydrocarbons is hydrogen sulphide (H_2S). A natural gas contaminated with this will cause pollution problems and is termed a 'sour' gas; a 'sweet' gas has less than one part in a million of hydrogen sulphide. Table 19 shows a range of typical constituents and calorific values of natural gas.

Natural gas can occur almost entirely on its own; dissolved under pressure in oil; or in a layer above the oil in a reservoir – a so-called 'gas-cap'. Natural gas is also found associated with coal, when it is a major hazard because it forms an explosive mixture with air. Under some conditions, however, it can be extracted from coal seams in useful amounts. Natural gas from coalmines is sometimes called colliery methane.

In the technical language of the oil industry, gas which occurs on its own is called 'non-associated', that occurring together with oil is called 'associated'. It has been estimated that about 72 per cent of the world's proved gas reserves are non-associated, 17 per cent are dissolved, and 11 per cent are found in gas-caps.

There is also a group of compounds which occurs at the lighter

Table 19. Constituents and calorific values of natural gas

Constituents	% of total volume		
HYDROCARBONS	'wet'	←(*range*)→	'dry'
methane	84·6		96·0
ethane	6·4		2·0
propane	5·3		0·6
iso-butane	1·2		0·18
N-butane	1·4		0·12
iso-pentane	0·4		0·14
N-pentane	0·2		0·06
hexanes	0·4		0·10
heptanes	0·1		0·80
NON-HYDROCARBONS			
carbon dioxide		0–5	
helium		0–0·5	
hydrogen sulphide		0–5	
nitrogen		0–10	
argon		less than 0·1% of helium content	
radon, krypton, xenon		traces	
Calorific values: 9·31–11·38 kWh/m³ (900–1 100 Btu/ft³)			

Source: E. N. Tiratsoo, *Oilfields of the World*, Scientific Press, 1973.

end of the petroleum oils. These are known as 'natural-gas liquids' or sometimes as 'natural gasoline' or 'condensate'. They are frequently found in conjunction with natural-gas deposits. They may occur as liquids or they may condense from the gas when it is released from the high pressure of the reservoir. The term 'wet' is applied to a gas in which the extractable liquid hydrocarbon content is more than 1 litre per 25 cubic metres of gas (0·3 gallons per 1 000 cubic feet).

Each mode of gas occurrence requires its own recovery technique. That for non-associated gas is relatively simple. If the gas is dry and sweet it can be piped almost directly from the well to the consumer with little more than pressure regulation controls. Wet or sour gas, however, must be processed to remove the liquids or contaminants. Non-associated gas is obviously not extracted unless or until there is some use for it.

The problems with associated gas are greater. Because a well producing such gas is usually operated for optimum oil recovery, the gas is obtained, essentially, as an unavoidable by-product. In the early days of the oil industry associated gas was regarded as an unmitigated nuisance and was either flared off or allowed to blow away. And it still causes difficulties. If no means of storing it or distributing it exists, it is usually flared. It can, however, under suitable conditions, be reinjected into the oilwell to maintain the pressure for oil extraction. In that case it remains, at least theoretically, available for future recovery.

Dissolved gas bubbles out of the oil when it reaches the surface and is released from the reservoir pressure. The amount of gas held in solution in an oil is called the 'gas–oil ratio' and varies from less than a hundred to several thousand cubic feet of gas per barrel of oil. The gas must be removed before the oil can be safely sent in a pipeline or loaded into a tanker. The decision whether or not to flare the gas depends on the quantity of gas and the economics of distributing or reinjecting it.

The use of gas requires a distribution system or a conveniently situated petro-chemical works which can use it as a feedstock. In the US these requirements are relatively easy to meet and there are legal penalties for wasting gas. Over the rest of the world, however, gas wastage is still appalling. The Middle East gas flares and the Great Wall of China were the first distinguishable man-made objects sighted by one of the returning moon missions.

In the Middle East and Africa over 60 per cent of the associated gas is flared. The total amount of gas flared in the world in 1978 was estimated to be about 154 billion cubic metres. This was about 9 per cent of the total natural gas produced and was the equivalent of about 130 million tonnes of oil – about the same as the total commercial energy consumption of the whole of Africa, or that of the whole Indian subcontinent with its population of over 800 million. The rush to extract oil as quickly as possible is costing the world dear in wasted energy. The major oil-producing countries are, however, now making considerable efforts to reduce the wastage of gas and are building distribution pipelines and industrial plants which can use it. One of the most important uses is the manufacture of ammonia for nitrogenous fertilizer production.

Table 20 gives the distribution of the world's present proved natural gas reserves. The most thorough exploration has been carried out in the United States, which is by far the largest producer and consumer of natural gas. In the oil-producing countries of the

Table 20. World proved natural gas reserves – selected countries, 1979

Country/region	Reserves cu. m $\times 10^9$	% of world total
ASIA–PACIFIC		
Australia	848	1·2
Indonesia	1 088	1·6
TOTAL: ASIA-PACIFIC	3 801	5·5
WEST EUROPE		
Netherlands	1 660	2·4
United Kingdom	705	1·0
TOTAL: WEST EUROPE	3 499	5·1
MIDDLE EAST		
Abu Dhabi	780	1·1
Iran	10 700	15·5
Iraq	790	1·1
Kuwait	1 120	1·6
Saudi Arabia	2 105	3·1
TOTAL: MIDDLE EAST	16 147	23·5
AFRICA		
Algeria	2 970	4·3
Libya	695	1·0
Nigeria	1 140	1·7
TOTAL: AFRICA	5 138	7·5
WESTERN HEMISPHERE		
Canada	1 871	2·7
Mexico	1 723	2·5
Venezuela	1 190	1·7
United States	5 716	8·3
TOTAL: WESTERN HEMISPHERE	11 508	16·7
COMMUNIST AREAS		
USSR	27 500	39·9
China	707	1·0
TOTAL: COMMUNIST AREAS	28 761	41·7
TOTAL WORLD	68 854	100·0

Source: *Jeffrey Segal*, 'Natural Gas: World Survey', *Petroleum Economist*, August 1979.

Note: Only countries with reserves over 500 cu. m$\times 10^9$ included in the above tables.

developing world there is usually no readily accessible market for gas so it has not been a major concern of the oil companies to prospect for and prove gas reserves. Discoveries have therefore tended to be made fortuitously in the search for oil.

The total reserves figure of 68 854 billion cubic metres is equivalent to about 59 billion tonnes of oil. Proved reserves are thus about two-thirds those of oil. The pre-eminence of the Middle East in oil reserves is not reflected in the picture for gas. The USSR has by far the largest amount, about 40 per cent of the total: a result of the extraordinary richness of Western Siberia. Iran has the next largest reserves with 15 per cent of the total, almost twice as much as the US.

In 1970 the proved reserves of the world were about 38 000 billion cubic metres; they have thus almost doubled over the past decade. During that time, by contrast, the proved reserves of the US, the Netherlands and the UK have fallen by about 30 per cent.

The estimation of the world's ultimately recoverable reserves of natural gas is subject to even greater uncertainties than that of oil. While numerous estimates have been made over the past twenty years no clear convergence on a single figure, as in the case of oil, is apparent. Hubbert, in 1973, suggested a figure of about 340 000 billion cubic metres, almost exactly equivalent to the figure of 2 000 billion barrels, for the world's ultimately recoverable oil reserves. In this instance Hubbert is among the optimists.

Another estimate made in 1978 by Meyerhoff suggests a figure of about 200 000 billion cubic metres. Details of this estimate are given in Table 21. This allows a comparison to be made between presently proved reserves and those possibly yet to be found. The table clearly reveals the lower prospects for the US when compared with the other less explored potentially gas-bearing areas of the world. Again, the predominant position of the USSR is to be noted. With almost half the world's ultimate natural gas reserves and an even higher proportion of the world's coal, the USSR is by far the most richly endowed country in the world.

Table 21. World natural gas reserves – discovered and ultimate – 1978 cu. m $\times 10^9$

Country/region	Discovered	Undiscovered	Total	% of Total
USA	6 148	2 265	8 413	4·2
Canada	2 768	9 911	12 679	6·4
Other Western Hemisphere	3 515	2 645	6 160	3·1
Western Europe	4 086	5 799	9 885	5·0
Iran	14 571	11 327	25 898	12·9
Other Middle East	5 761	11 392	17 153	8·6
Africa	5 321	4 740	10 061	5·0
Asia/Pacific	3 593	9 251	12 844	6·5
USSR	25 499	61 731	87 230	44·0
China	816	7 702	8 518	4·3
TOTAL	72 078	126 763	198 841	

Source: A. A. Meyerhoff, 'Proved and Ultimate Reserves of Natural Gas and Natural Gas Liquids in the World', quoted in *Petroleum Economist*, December 1979.

No allowance is made in the table for possible gas reserves obtainable from deep formations in the 4 000–5 000-metre range, from 'tight sands' and zones containing high-pressure brine solutions. These have generally been considered uneconomic or beyond the abilities of presently available technology. But rising prices and technical advances are making them seem more attractive and some commentators believe they may turn out to be extremely rich sources in the future. No reliable assessments of the amounts of gas which might be obtained from these new sources are yet available.

Table 22 shows the world consumption and production of natural gas. These have been growing at about 3½ per cent per year since the early 1970s. Natural gas now supplies nearly a fifth of the world's total energy and just over a quarter of the energy consumed in the US and the USSR. Sixty-nine countries now produce natural gas commercially, with Ireland being the latest. The US is still the largest producer and consumer but production has been

Table 22. World natural gas consumption and production – 1978

cu. m × 10^9

Country/area	Consumption	Production
United States	588	557
Canada	55	72
TOTAL: NORTH AMERICA	643	629
Latin America	49	61*
France	24	8
Italy	28	13
Netherlands	40	90
United Kingdom	44	39
W. Germany	49	20
TOTAL: WESTERN EUROPE	209	186
Africa	10	40
Middle East	30	40†
Japan	20	3
USSR	337	372
East Europe	76	56
China	39	50
TOTAL: WORLD	1 448	1 458

†Of which Iran 22 *Of which Mexico 31

Sources: Consumption: *BP Statistical Review of the World Oil Industry.*
Production: *Petroleum Economist*, August 1979.

Note: Only countries and regions with more than 20 × 10^9 cu. m consumption or production included in the above table.

falling in recent years; the fall has been partly compensated by imports, mainly from Canada. In the USSR production and consumption have both been increasing.

Unlike oil, most of the world's natural gas is consumed in the countries in which it is produced. The reason for this is obvious. Natural gas is almost ideal as a fuel: it has a high calorific value, it is clean and efficient in use, and the only products of its combustion are carbon dioxide and water. But it is difficult to transport and distribute. For large-scale use it requires a network of underground pipes connected to every consumer. To establish such a system from scratch is a long and expensive undertaking. It can only be justified if a country has its gas supplies under its own control or is confident of the goodwill and political stability of potential suppliers and the countries through which supply pipelines must pass.

The alternative to pipelines for large-scale transport is the use of refrigerated tankers which transport liquefied natural gas (LNG). These are expensive, complicated and dangerous. They also require elaborate liquefaction and regasification facilities. There is controversy about the possibility of escaped liquid natural gas mixing with air to form an explosive mixture which could ignite from a spark. It has been argued that the gas would quickly disperse to safe concentrations and that such an explosion on a large scale is impossible. The fact that a major explosion of gas in the open air is indeed possible was tragically demonstrated in Flixborough in England in June 1974 when a chemical works and a nearby village were devastated with a loss of twenty-nine lives. The energy content of a large LNG tanker is as great as that of a major thermonuclear weapon. A collision in a port which caused a tanker's load to spill and explode is potentially as damaging as a nuclear bomb explosion. It is astonishing that there has been so little public concern about this.

In future, and in spite of the dangers, it is probable that more and more LNG will be transported. The fall-off in production in the US and Europe will to some extent be offset by such imports. The US is anxious to increase its imports by pipeline from Canada and Mexico, and by LNG tanker from Indonesia and other countries. It is unlikely, however, that any suppliers will be able to compensate for the decline in domestic US reserves. In Europe the largest producer, the Netherlands, is making arrangements for imports from Algeria and Norway, and a second trans-Mediterranean submarine pipeline is being considered. Such measures may well be sufficient to maintain European consumption of gas at present levels for at least the next few decades.

Another possibility is that western Europe might be linked to the USSR's resources in a major way. One pipeline already runs from the USSR through Western Germany to the centre of France.

This could be the beginning of a wider network bringing sufficient gas to western Europe to meet its needs for the next century. Some delicate political problems would naturally have to be solved first.

Countries with large reserves, notably the USSR, but also in the Middle East and China, will undoubtedly greatly increase their own consumption. In the oil-producing countries which have now taken control of their resources, the wastage of gas will also be cut back and its productive use increased. This, however, may not add much to the energy entering world markets. The arguments against exporting more oil than necessary apply with equal force to gas. Increased exports of gas could even lead to a reduction in oil exports.

The long-range importance of natural gas could, however, be more important if some recent speculations about the non-organic origin of some natural gas and its widespread presence deep in the earth's crust are soundly based. The presence of methane in volcanic eruptions and some earthquakes suggests it may be coming from greater depths and sources other than the sedimentary basins. In theory methane could be synthesized in large amounts from non-organically derived hydrogen and carbon. It remains to be seen if in fact it is, and what effect, if any, this will have on future supplies of natural gas.

9

Alternative Sources of Oil

Tar Sands and Heavy Oils

Sands or sandstones impregnated with heavy oils occur in a number of areas in the world. The largest deposit is near Fort McMurray on the Athabasca River in the Canadian province of Alberta. It covers an area of about 34 000 square kilometres, about the size of Belgium. In eastern Venezuela there is a 2 300-square-kilometre deposit of somewhat lighter hydrocarbons in the Oficina–Temblador area. There is a further huge deposit at Olenek in the USSR. Although there are a score or more other large deposits around the world none approaches these three in size.

Most of the development work so far has taken place on the Athabasca deposit, since economically and technically this is more promising than any other. The thickness of the tar sands there is between 40 and 80 metres, with an average of about 55 metres. In some places the tar sands are found at the surface, notably along the Athabasca River where there are cliffs of the material up to 45 metres high. More usually there is an overburden of glacial material of about 80 metres, though this increases to as much as 600 metres in some areas.

The precise origin of the tar sands is still a puzzle to geologists. Some believe the bituminous material was formed where it now lies; others believe it migrated there from the underlying Devonian beds or downwards from the now mainly eroded Cretaceous deposits. The original petroleum may have resembled a conventional crude oil but almost all the volatile components have been lost. The tar sand is a thick heavy material with a sticky texture. In the ground it is interspersed with layers and lenses of sands, clays and shales. It is an awkward material to extract and process.

The presence of the tar sands in Athabasca has been known for a long time and many attempts have been made to extract the tar or to develop uses for the material. As early as 1915 a demonstration road surface, using the mined material as a kind of natural macadam, was laid and performed satisfactorily. But this and other ventures ended in commercial failure. The Athabasca region is very wild and desolate, with an appalling winter climate. The ground cover is mostly muskeg, a soft peaty substance which makes transport exceedingly difficult. S. C. Ells, one of the pioneers who spent his life exploring the area and trying to develop ways of using the tar sands, described the extreme discomfort caused by the myriads of flies with which the area is infested.[24]

It was only during the late 1960s that the technical problems of separating the oil from the sand were solved on a commercial basis by a company called Great Canadian Oil Sands which began work in 1967. The project ran at a heavy loss during the first five years but then showed sufficient promise to encourage further developments by other countries. Output from the Great Canadian Oil Sands plant is now reported to be about 50 000 barrels per day. Another project, which has been heavily supported by the Canadian Federal and Provincial Governments is expected to commence operation, with an output of 15 000 barrels per day in 1980. Plans are being made for further increases and within a decade or so Canada might be obtaining up to half its oil from tar sands.

As about 2 tonnes of mined material are required to produce a barrel of oil, the scale of operation necessary for an output of oil comparable with that from a conventional oilfield is gigantic. First the overburden has to be stripped and dumped. Next draglines or bucket-wheel excavators have to dig out the material which is then taken to the separation plant. For the operation mentioned above this means digging out 250 000 tonnes of material every day and disposing of the same amount of waste.

At the separation plant the raw sand is mixed with hot water and injected with steam to bring the mixture to a temperature of about 80°C. A series of filters and foam-raking processes then separates about 90 per cent of the bituminous material which is taken for refining in an almost conventional way. One of the problems is that the bitumen has a high sulphur content and also contains metallic pollutants. The hot waste sand is dumped back into the excavated area of the mine and the process water purified and recycled.

The mining operation is similar to strip-mining coal. The operating conditions, however, are considerably worse than those usually encountered in coal mining. The freezing of the tar sands to a rock-like substance in winter has caused numerous problems. Large tracts of land will be ruined but northern Alberta is thinly populated and it has not been suggested that environmental concern will hinder operations. The limiting factors seem likely to be in the technology of digging big holes and processing massive quantities of material.

Under the stimulus of the rise in world oil prices, development work on the deposits in Venezuela and the USSR has been accelerated. In Venezuela the target is 250 000 barrels per day by 1990 and an active programme is reported to be under way in the USSR.

World resources in place are estimated to be about 300 billion tonnes. About 90 per cent of these are divided equally between the three major deposits. Only 5–10 per cent of this quantity is accessible from the surface and hence recoverable with presently available methods.

No way of recovering the remainder yet exists. Work is being done to devise ways of softening the deposits *in situ* by pumping steam down boreholes or by lighting fires and feeding them with oxygen from the surface. These processes, some of which are used in the tertiary recovery of crude oil, might sufficiently reduce the viscosity of the oil to enable it to be recovered by pumping from boreholes. Perhaps a further 10 per cent of the deposits might be recoverable in this way. The total recoverable reserves in the tar sands and heavy oils thus amount to perhaps 50 billion tonnes, about the same quantity as the known reserves of the Middle East. This is a large quantity of oil but until *in situ* methods of recovery are developed all but a small proportion of it remains completely inaccessible.

Oil Shales

Oil shale is a relatively common, finely textured sedimentary rock containing the solid organic material kerogen. The oil shales were formed in many different geological periods and hence vary considerably in composition and richness. They are found in many parts of the world. On heating to 300–400°C, in a retort or large distillation vessel, the kerogen breaks down into a number of gaseous and liquid hydrocarbons which can then be extracted. The liquids resemble a crude oil in some respects, but cannot be used directly in a conventional refinery: a preliminary upgrading process is required.

In the seventeenth century oil distilled from shale was used in the Modena district of Italy to provide street lighting. France had an oil shale industry as long ago as 1838. In Scotland shale oil was produced from 1848 until it became uneconomic to do so in 1962. Production was over 100 000 tonnes a year in the early 1950s. Pulverized oil shale is used as a direct fuel for electric-power generation in Estonia, and in the manufacture of gas for Leningrad. There is also a large oil shale industry in China in the Funshun province of Manchuria, where oil production is apparently in millions of tonnes a year.

The richest deposits of oil shale are found in the United States, in the Green River area of Colorado, Utah and Wyoming. About 44 000 square kilometres here are underlain by oil-bearing shales. Within this area the richest deposits are concentrated in the Piceance Creek Basin (spelled 'Pissants' on the maps of a more robust age) which covers about 440 square kilometres and contains 80 per cent of the potentially recoverable oil in the area. The quantity of oil contained in these shales has given rise to some exaggerated optimism about their potential as a 'solution' to the world's energy problems.

This 'solution' remains to be found. The organic content of the Colorado oil shales varies from about 2 barrels per tonne down to near zero. The richest deposits are up to a hundred metres thick but these are generally deeply buried. Much of the oil is contained in thin layers interspersed with rocks of low or zero oil content. Although the oil shales outcrop along river valleys and some other areas, they cannot usually be mined by open-cast methods. In contrast with the tar sands the oil shales are hard rocks. Shale for retorting has to be quarried out of deep workings or mined in much the same way as coal. Only the richest deposits are worth working in this way. A tonne of coal is a tonne of coal. But a tonne of oil shale is a lot of rock and a little oil which has to be retorted and upgraded. Obtaining the same amount of energy from an oil shale which yields a barrel of oil per tonne requires five times as much mining as coal. Mining can, in fact, account for 60 per cent of the cost of oil from shale. It also causes big problems of environmental despoliation and waste disposal. For these reasons there is intense eagerness to develop a method of retorting the oil shales *in situ*.

Essentially this would have to duplicate the above-ground process. The shale must be fractured, heated in a controlled manner, and the liquid and gas products drawn off. There are enough problems when these operations are conducted on a large scale above ground; when they are out of sight a hundred metres below ground, the difficulties of control are quite daunting. For fracturing the rock, explosives, hydraulic pressure, and high-voltage electricity have all been suggested. Combustion would be fed with air or oxygen from the surface. A method of keeping the temperature within the tolerances required by the retorting process is essential if the oil produced is not to vary unacceptably in quality and composition.

One apparently promising process has been developed in a pilot plant by the Occidental Oil Company. Tunnels are driven horizontally inwards from a cliff face. A large chamber of fractured shale is then created with explosives. Natural gas is piped in to start a fire which is fed with oxygen. Oil dis-associated by the heat from the shale flows to the bottom of the combustion chamber where it collects in a sump from which it can be drawn off.

Inevitably, the use of nuclear explosions has been suggested to fracture the shale and provide sufficient heat to retort it. There are, however, difficulties in predicting the effects of underground nuclear explosions in detail. It also seems impossible to prevent the oil from becoming radioactive, which would make it unusable. As in the case of tertiary recovery from oil wells, the main supporters of this seem to be people with interests in the manufacture of nuclear bombs – for which, admittedly, it is difficult to find other productive uses.

Not surprisingly there is confusion about the size of oil-shale resources. Measured as 'oil in position' they are immense. A calculation of the total contained in all the world's deposits containing 5 gallons of oil per ton and over is quoted by Hubbert. It is 2×10^{15} barrels, a thousand times greater than the world's recoverable conventional crude oil. But again, the crucial question is how much of this is recoverable.

Table 23 shows the distribution of the world's shale-oil deposits together with estimates of their oil content. The figures are indicative only; little work has been done to identify or evaluate shale-oil resources throughout the world because on the whole they have not been economically attractive except in limited areas. It is now felt that only those shales with an oil content above 25 gallons per tonne will ever be economic. This reduces the size of the potential reserves by a factor of a thousand. At the 1978 World Energy Conference it was estimated that the recoverable oil from these resources was about 1 500 billion barrels – about 90 per cent of it in the US. The total is thus somewhat less than conventional crude oil. And of that only 5–10 per cent can 'be considered for immediate exploitation.'

In fact, shale oil does not look at all promising as an energy resource. The environmental problems are being viewed with

Table 23. Estimated shale-oil resources of the world

	barrels × 10^9		
	Total resources in position Oil content, gallons/ton		
Continent	5–10	10–25	25–100
Africa	450 000	80 000	4 000
Asia	590 000	110 000	5 500
Australia & New Zealand	100 000	20 000	1 000
Europe	140 000	26 000	1 400
North America	260 000	50 000	3 000
South America	210 000	40 000	2 000
TOTAL	1 175 000	325 000	17 000

Source: Duncan & Swanson, 1965 – quoted by M. K. Hubbert in *Resources and Man*, W. H. Freeman, 1969.

increasing pessimism. The lack of water in the arid areas of the US in which the richest deposits are located could well be the greatest constraint of all. Some oil industry experts believe this could limit the production of oil from the Colorado shale to a maximum of 50–100 million tonnes a year. Hubbert is even more pessimistic. His comment is that '. . . the organic contents of the carbonaceous shales appear to be more promising as a resource of raw materials for the chemical industry than as a major source of industrial energy'.[17]

One of the most telling factors is the sheer magnitude of the shale-oil mining and quarrying operations necessary to produce yields comparable with those from coal or oil. If half the present US oil consumption were to be supplied from shales yielding a net quantity of 30 gallons per ton then the total amount of shale mined would amount to 4¼ billion tons per year. This is about one and a half times the output of the whole world's coal-mining industry. It is difficult to see how this could actually fit into the Piceance Creek area, quite apart from any question of whether it would be socially or environmentally acceptable there.

10

Nuclear Energy

Fission Power

In a nuclear power station atoms of the radioactive element uranium undergo a process of fission, disintegrating into lighter elements and sub-atomic particles. If the products of this disintegration are added together they do not quite add up to the original mass of the uranium atom. The mass which has vanished has been transformed into energy. The energy can be calculated from Einstein's equation $E = mc^2$; energy is equal to the mass 'consumed' multiplied by the velocity of light squared. For a single atom this mass is very small, but c^2 is very large (the velocity of light is 297 600 kilometres per second) so that on the scale of a nuclear reactor, or a bomb, the amount of energy produced is prodigious.

An atom can be visualized as a miniature solar system in which electrons orbit about a compact nucleus of protons and neutrons. Most of the mass of the atom is concentrated in the nucleus; the mass of a neutron or a proton is about 1 800 times that of the electron. The proton has a positive electrical charge; the electron has a negative charge of the same magnitude. Since atoms have no net electrical charge the number of protons must equal the number of electrons. The neutron, as its name suggests, has no charge.

Atoms with the same number of electrons are chemically indistinguishable. The simplest atom is hydrogen which has one electron; the most complex naturally occurring atom is uranium with ninety-two electrons. The numbers of electrons, or protons, in a single atom of an element is called its 'Atomic Number'.

The mass of an atom, misleadingly called its 'Atomic Weight', is the total mass of the protons, neutrons and electrons of which it is composed. If the small mass of the electrons is ignored, the Atomic Weight can be expressed as a whole number, representing the number of protons and neutrons contained in the nucleus of the atom. The Atomic Weight of hydrogen, which has a single proton in its nucleus is therefore 1; the Atomic Weight of helium, the next simplest atom, is 4 because it has a nucleus of two protons and two neutrons; that of carbon is 12 because it has a nucleus of six protons and six neutrons.

Following on from his work with Rutherford at the turn of the century Soddy discovered in 1913, however, that some atoms which are chemically the same have different masses. These Soddy christened 'isotopes', a word which means literally 'same place', because they occupy the same place in the Periodic Table of the elements. The difference in mass occurs because they have a different number of neutrons in their nucleus.

Hydrogen has two naturally occurring isotopes: 'normal' hydrogen and deuterium. Normal hydrogen has a single proton in its nucleus; deuterium has a proton and a neutron. Chemically they are identical but the mass of deuterium is twice that of hydrogen. Uranium has three naturally occurring isotopes: uranium-238 which has 92 protons and 146 neutrons; uranium-235 with 92 protons and 143 neutrons; uranium-234 with 92 protons and 142 neutrons. Most elements occur in a number of isotopic forms.

Some isotopes are radioactive. The balance of neutrons and protons in the nucleus is unstable. These isotopes move towards a stable state by emitting energy or some of the sub-atomic particles of which they are composed. Radioactive decay is a complex phenomenon. The simple model of the atom composed of hard indestructible particles, protons, electrons and neutrons, does not fully explain what happens. In the radioactive decay of the atom some of the particles of which it is composed can themselves break down; the neutron may emit an electron and turn into a proton.

The early researchers, without clearly understanding what was happening, identified three kinds of 'emanations' or 'rays' emitted by radioactive substances. These they christened alpha-, beta- and gamma-rays. Subsequent investigations showed these to be fundamentally different and scarcely deserving the common designation of 'ray'. Nevertheless the names have stuck and are commonly used in discussions of radioactivity.

Alpha-rays are streams of heavy composite particles consisting of two protons and two neutrons. These particles are identical with the nucleus of the helium atom. Beta-rays are streams of electrons. Gamma-rays are true electromagnetic radiation. They have shorter wavelengths than X-rays and are extremely penetrating.

As a result of losing neutrons, protons, or both, or in the transformation of a neutron into a proton, an atom changes its identity. It can become a different isotope of the same material or it can become a different material altogether. In radioactive decay a genuine transmutation of elements can and does take place. The process of reaching a state of stability through radioactive decay may be a long one. An atom decays and becomes something else which may itself be unstable; another decay takes place, and so on until stability is achieved. In the decay of the naturally occurring radioactive element uranium-238 there are nineteen separate decay stages (in one of which radium is formed) before the final formation of a stable isotope of lead.

Although it is impossible to predict the moment when any individual atom will decay, the rate of decay of a large number of atoms is totally predictable. For every radioactive material the time in which half its atoms will decay can be measured. This is an unvarying amount. It is called the 'half-life' of the material. If the half-life of a radioactive substance is a hundred years, then half the atoms in any quantity of it will decay in a hundred years, half the remainder in the next hundred years and so on. The half-lives of different radioactive substances vary from billionths of a second to billions of years. All radioactive substances are energy sources. In most cases, however, the rate of energy emission is far too slow for any practical use – though miniature devices are sometimes powered by radium.

Uranium-235 is a naturally occurring radioactive isotope with a long half-life of $7{\cdot}1 \times 10^8$ years. In nature it occurs in the proportion of 1 part in 140 with uranium-238 which is also radioactive but with a much longer half-life of $4\frac{1}{2}$ billion years. (The third natural uranium isotope, uranium-234, occurs in negligible amounts.) Uranium-235 normally decays by emitting an alpha-particle. If this were all, there would be no nuclear bombs and no nuclear power industry. The unique property of uranium-235, among naturally occurring materials, is that its atoms occasionally split spontaneously into two approximately equal portions and emit two or three neutrons, as well as releasing a considerable amount of energy. This is nuclear fission. It does not, however, occur sufficiently frequently to give useful amounts of energy.

The nuclear industry is possible because uranium-235 can be made to split by hitting its nucleus with a slowly moving neutron. Thus, if the neutrons emitted in a spontaneous fission can be controlled in speed and directed into the nucleus of other uranium-235 atoms these will in turn split and release neutrons to keep the process going. Since each fission absorbs one neutron and releases two or three, if all the released neutrons are used to cause further fissions, the process will rapidly accelerate. This is the chain reaction which is the source of power for the atomic bomb.

The main problem in making a bomb is ensuring that most of the neutrons are used to cause further fissions. They must not escape from the mass of uranium; neither must they be absorbed by other substances. For the bomb to be 'efficient' the chain reaction should build up as rapidly as possible and create a massive blast-wave of heat and pressure. To achieve these desired effects a certain minimum quantity of uranium, called the 'critical mass', is required. With a smaller quantity of uranium the ratio of surface area to volume is too large and so many neutrons escape from the mass without causing a fission that the chain reaction cannot occur. To eliminate the problem of neutron absorption a bomb must use almost pure uranium-235, which has to be separated from the uranium-238 with which it naturally occurs. This process of separation is usually called 'enrichment'. A nuclear reactor, in distinction from a bomb, usually uses natural or slightly enriched uranium.

The requirements of a reactor are obviously different from those of a bomb. A reactor is designed to use a steady chain reaction to produce heat, which is then used to produce steam which in turn drives electricity-generating plant. In a nuclear power station the reactor plays the part of the coal- or oil-fired boiler in a conventional power station. (See Figure 9.)

Neutrons emitted from the fission of uranium-235 have different speeds but only those travelling slowly will cause other uranium-235 atoms to split. In a nuclear reactor a substance called a moderator, which slows neutrons without absorbing them, is used to bring some of the faster moving neutrons down to the speed at which they can cause a fission. In this way neutrons which would otherwise have been wasted become 'productive'. When each fission produces, on average, one productive neutron the rate at which nuclear fission is occurring is neither increasing nor decreasing. This steady chain reaction will continue as long as sufficient uranium-235 remains. The moderator, because its action brings the use of neutrons up to the required efficiency, is an essential ingredient of the system.

Water is a reasonably good moderator and so is carbon. One of the best moderators is heavy water. This is water in which the hydrogen atoms are the hydrogen isotope deuterium. It occurs in nature in the ratio of about one part in 6 500 in ordinary or 'light' water. Its separation, because there is no chemical difference between it and light water, is complicated and expensive, as well as being very energy-consuming.

Fine regulation of reactor operation is obtained by the use of control rods. These are made of a good neutron absorber such as cadmium or boron. When inserted into a reactor core they absorb neutrons and thus slow down the rate at which induced fission occurs. They can therefore be used to bring a reactor off power. In some reactor designs they are suspended by electromagnets above

Figure 9. Operation of conventional and nuclear power stations – schematic

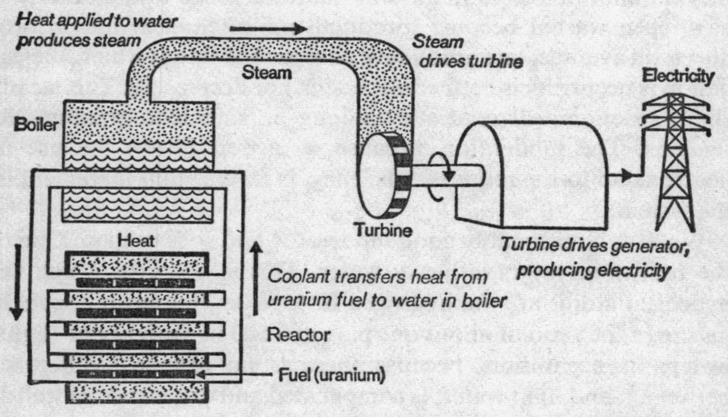

the core. In the event of any malfunctioning which cuts off the power supply to the electromagnets the control rods automatically drop into position and shut down the reactor.

When a nuclear reactor is operating it produces heat. Whether it is usefully used or not, this heat must be removed otherwise the reactor would melt. This is done by a 'coolant' circulating through the reactor core. A coolant should have a high heat-absorbing capacity; it should not, on the other hand, absorb too many neutrons, it should be non-corrosive, and it should be stable within the operating temperature range of the reactor. There is a variety of coolants. Gas-cooled reactors use air, carbon dioxide or helium. Liquid-cooled reactors use water, heavy water, liquid sodium or a molten alloy of sodium and potassium. In some present reactor designs the coolant is circulated directly through the electricity-generating turbines. In most cases, however, an indirect system is used and the coolant passes through a heat exchanger where it gives up its heat to produce the steam which then drives the turbines as shown in Figure 9. There is a bewildering variety of reactor types using different combinations of coolant, moderator and fuel type. They are usually referred to by sets of initials or technical nick-names.

The earliest nuclear reactors to provide commercially available electricity were built in Britain – the first was opened at Calder Hall, Cumberland, in October 1956. This and the next British 'commercial' station were primarily designed for the production of plutonium for military purposes. Thereafter, the same basic design principles were used in the next nine British stations. They were all called Magnox stations because this is the name of the special magnesium alloy used to clad the uranium fuel rods in the reactor. The early reactors were contained in a welded steel pressure vessel; the later ones in one made of prestressed concrete. In all, nine Magnox stations were built for commercial operation by the electricity boards in Britain; a further one was built in Japan, and another in Italy. Although there have been some operating problems and corrosion of reactor parts has led to a reduction in the core temperature with a consequent loss of efficiency, the Magnox reactors have generally performed well. The design, however, has inherent limitations and in order to overcome these a new approach was sought by British designers.

The only other country to embark on a programme of gas-cooled reactors was France. Two small units, one at Marcoule, the other at Avoine, started operation in 1958. In all, France built seven gas-cooled reactors, but by 1970 interest in this type of reactor had ended.

British designers persevered, however, and the Second Nuclear Programme announced in 1964 was based on a new higher-temperature gas-cooled reactor, called the Advanced Gas-cooled Reactor (AGR). The coolant is again carbon dioxide and the moderator graphite. The coolant gas comes from the reactor at about 650°C – about 300°C higher than the Magnox stations; the fuel is slightly enriched uranium, clad in stainless steel. While there are undoubted theoretical advantages in the AGR design the programme has been an almost unmitigated disaster for the British nuclear industry. Design changes and delays have led to the bankruptcy of contractors. Of the five large stations ordered, only two have come on stream – one of which, Hunterston, promptly disabled one of its reactors by sucking sea-water into the reactor pressure vessel. Dungeness B has become a nightmare; originally scheduled to come on stream in 1970, it is still in the throes of construction in 1980. The bright hopes of the British nuclear industry, which for almost a decade seemed to have a firm place in the forefront of nuclear developments, have been almost completely destroyed in the AGR programme.

A further development of the gas-cooled reactor is the High-Temperature Gas-cooled Reactor (HTGR). This operates at a temperature of at least 1 000°C. If it could be brought into commercial use it would open up a new range of uses for nuclear power as a source of high-temperature heat – far higher than anything which can be provided by present commercial reactor types. At this temperature, however, metallic fuel elements cannot be used and ceramic materials are employed instead. The fuel is highly enriched uranium in the form of pellets which are baked into a shell of silicon dioxide. Helium is used as the coolant and graphite as the moderator. A small reactor, called Dragon, was built and operated from 1966 until it was closed in 1976 as a joint international experiment under the auspices of OECD at Winfrith in Dorsetshire. Other small experimental units were built in Germany and the US and a

300-megawatt unit is expected to come into operation in Germany in 1982. The only commercial prototype to come into operation is at Port St Vrain in Colorado. This is a 330-megawatt unit which began operation in 1979. Orders for a series of large stations based on this design were placed some years ago but have all since been cancelled. Although it has many interesting and attractive features, the HTGR remains in a technological limbo at the moment.

By far the most common reactor now being built is the US light-water reactor (LWR). This uses ordinary water as a coolant and moderator; the fuel is slightly enriched uranium. Two varieties of the type have emerged: the pressurized water reactor (PWR) and the boiling water reactor (BWR).

The PWR operates at a pressure of 150 atmospheres. This is sufficient to prevent the water boiling at its operating temperature of about 300°C. The reactor is housed in a welded steel pressure vessel. The high operating pressure places extremely heavy demands on the quality of fabrication of the pressure vessel and other pressurized pipes and components. It is believed this was one of the reasons the UK government rejected the PWR design for the next phase of the UK nuclear programme in the early 1970s.

The BWR is the simplest reactor of all. It operates at a pressure of about 70 atmospheres and a temperature of about 300°C; under these conditions the water turns into steam. It was originally thought that water boiling within the reactor core would make control impossible but this has not turned out to be the case.

Both designs were derived from the reactors used in the US nuclear submarine programme. The first PWR began operation in Shippingport near Pennsylvania at the end of 1957; the first BWR came on stream a little over a year later near Chicago. These two reactor types dominate the market for nuclear power. Westinghouse are the manufacturers of the PWR and General Electric the manufacturers of the BWR. Virtually all the reactors now being built outside the Communist countries are either under licence to these two companies or being built directly by them. A substantial part of the USSR nuclear programme is also based on a version of the PWR. In a survey conducted in early 1980 it was found that of the world's total number of reactors operable, under construction or on order, 289 were of the PWR type, 117 of the BWR type and the remaining 124 were divided between a dozen different types some of which have, like the Magnox, been superseded or which have very limited commercial possibilities, such as the AGR.

A Canadian system called CANDU (Canadian deuterium uranium) uses heavy water as a moderator and coolant. Instead of the whole reactor being enclosed in a pressure vessel, this system uses a series of pressure tubes which contain the fuel and through which the coolant passes. The pressure inside the tubes is about 90 atmospheres. The tubes run through a large tank filled with heavy water which acts as the moderator. The first CANDU station came on stream in Pickering, Ontario, in 1971. The system is still the basis of the Canadian nuclear programme. CANDU stations have been exported to the Argentine, India and Pakistan, but the Indian explosion of a nuclear bomb has caused Canada to review its export of nuclear technology to developing countries. The CANDU reactor has good safety characteristics and it has proved reliable in operation but it has not been able to compete commercially with light-water reactors in world markets.

In the USSR a graphite-moderated, light-water-cooled reactor came into commercial operation in 1958. Considerable development of this system has taken place and units of up to 1 500 megawatts are under construction.

Numerous other reactor designs have been attempted. Early efforts in Switzerland and Sweden to develop indigenous designs came to nothing. A British design called the steam-generating heavy-water reactor (SGHWR) was based on the same principles as the CANDU but used light water as a coolant. This was allowed to boil and was then used to drive the turbines directly, as in the BWR. The SGHWR was briefly considered as a candidate for commercial development by the UK government in the middle 1970s but has now been abandoned.

The breeder reactor is designed to exploit to the fullest possible extent one of the important changes which occurs in the fuel rods of all reactors as the reaction proceeds. When the uranium-235 atom splits into two parts, substances such as strontium, xenon, krypton and others are formed. These are called 'fission products'. They absorb neutrons and are therefore said to 'poison' the fuel because they gradually reduce its energy output. After a time, which varies between reactor systems, the fuel must be removed because the depletion of uranium-235 and the build-up of poisoning make it impossible to sustain the nuclear reaction.

Not all neutron absorption, however, is detrimental to the reaction. One of the substances which absorbs neutrons is uranium-238, the principal constituent of the fuel elements. When uranium-238 absorbs a neutron it decays radioactively and becomes plutonium-239. This is a radioactive substance not found in nature; it has a half-life of 24 000 years; it is also a bone-seeking poison. It is possibly the least desirable substance man has added to his planet. But some ruthless experiments in the United States during the late 1940s in which terminally ill cancer patients were given massive doses of plutonium belie the most horrific elements in its reputation: some of the patients survived for thirty years. It has the property of being fissionable by neutrons in the same way as uranium-235; and it has the advantage of being fissionable by fast as well as slow neutrons. It is therefore usable as a fuel in nuclear reactors or atomic weapons. In fact, every operating reactor produces and uses some plutonium as it depletes its uranium-235. In the CANDU stations up to half the total energy output is derived from the fission of plutonium created within the reactor.

The breeder-reactor concept takes this logic further. In a stable reaction each fission need only produce one 'productive' neutron which causes another fission. This uses less than half the total number of neutrons produced. If all the remainder could be employed to change uranium-238 into plutonium then the amount of plutonium would be greater than the amount of uranium-235 consumed. There is nothing magical about this; the fire is not producing more fuel than it consumes. Neutrons which would otherwise go to waste are simply being used to unlock the energy in the uranium-238.

In the breeder reactor, therefore, a major effort is made to avoid neutron waste. Heavily enriched fuel, up to 75 per cent uranium-235, or plutonium, is used in a very compact core without any moderator. Around the core is placed a 'blanket' of uranium-238 which absorbs neutrons and 'breeds' new plutonium. The compactness of the core and the intensity of fission activity there cause major cooling problems. A high heat-capacity material, moving very quickly, must be employed. This is why liquid metals, pure sodium or a sodium-and-potassium alloy, have been used as coolants in the breeder reactors built to date. But because these liquid metals, in passing through the reactor core become radioactive, an additional, intermediate heat-exchange system, also using liquid metal, is required before the heat can be used to generate steam for the turbines.

The attraction of the breeder reactor is that it makes use of the uranium-238 which for the most part remains unused in other reactors. Since uranium-238 constitutes by far the greater proportion of the world's uranium the increase in effective energy resources is very large. In theory, breeder reactors improve the utilization of uranium resources by a factor of 100. If they were used entirely in breeder reactors the world's uranium resources would provide a virtually inexhaustible supply of energy.

Breeder reactors are not new. The world's first nuclear electricity, albeit a tiny amount, came from a US breeder reactor, the EBR1, in 1951; four years later, however, its core melted. The first true power reactors were the British Dounreay Fast Reactor (DFR) which came on stream in 1959 and the Edison Fermi near Detroit. The DFR performed satisfactorily until it was closed in 1977; the Edison Fermi had a dramatic history including a partial melt-down of the core in 1966, during which, it was reported, the evacuation of Detroit was seriously considered.

Over the past decade, however, progress has been rapid. The French Phenix of 350 megawatts came on stream in 1973; in the same year the USSR commissioned its Shevchenko plant, which has a capacity of 350 megawatts and produces both electricity and distilled water. Phenix has operated satisfactorily; there were early rumours of an explosion associated with the Shevchenko plant, but whatever difficulties arose have apparently been overcome. In the UK the Dounreay Prototype Fast Reactor (PFR) of 250 megawatts came on stream in 1974. In both Japan and Germany 300-megawatt plants are expected to come on stream in 1985. France has taken the biggest step beyond this and Super Phenix, a station of 1 200 megawatts, is expected to be completed by 1983. In the USSR the 600-megawatt station at Beloyarsk is expected to come on stream in 1980.

At present, breeder reactor technology is being vigorously developed in the major industrial countries with the notable exception of the United States. The publicity surrounding President Carter's announcement that the US was withdrawing from breeder reactor development until there should be a means of controlling the proliferation of nuclear weapons (which could be made from the plutonium produced by breeder reactors) concealed the disarray in which the US breeder reactor programme found itself. There had been bitter disagreements both about the direction of the programme and the shares which should be borne by government and private industry. The debate on the breeder reactor in the US is far from over.

The term 'fast' is sometimes used when referring to breeder reactors. This has nothing to do with the speed at which plutonium is produced. This is, in fact, rather slow. Even under optimum conditions, breeders built to present designs produce only about 10 per cent more plutonium than they consume in any year. Building up a stock of plutonium sufficient to start a new station would therefore take a minimum of ten years – and, in practice, this is much more likely to be twenty years or more. The term 'fast' refers to the use which is made of fast neutrons. Other reactors are sometimes referred to as 'thermal' because they can only use neutrons which are travelling at slow speeds compatible with the thermal energy of the reactor core.

Thorium-232 can be mentioned here. It is a naturally occurring radioactive element with the very long half-life of 14 billion years. If it is exposed to neutron bombardment it changes into uranium-233. This is an element which does not occur naturally but which, like plutonium, behaves in the same way as uranium-235. It can therefore be used as a nuclear fuel. Uranium-238 and thorium-232, while they are not themselves fissionable, are capable of being transformed to fissionable materials: they are therefore termed 'fertile'. Both are 'second generation' nuclear fuels. The use of thorium at present is confined to limited experimental work in connection with the design of high-temperature gas-cooled reactors.

The spread of reactors currently operating outside the Communist world is shown in Table 24 together with their acronyms, moderators and coolants. They all have this imperative in common: their radioactive processes must be completely and at all times isolated from the outside world.

All radiation is dangerous. Alpha-rays, when they encounter another substance, crash through its atoms, quickly losing their own energy as they do so. They are not very penetrating; they are stopped by a piece of paper or the skin. But if they are released inside a living cell their short passage can be devastatingly disruptive. Beta-rays are more penetrating, and gamma-rays need very thick concrete or lead shielding to stop them. The main effect of beta- and gamma-rays is ionization, knocking electrons out of their parent atoms, thus altering their structure and behaviour. These rays are very damaging to living creatures.

Radiation damage shows itself in two ways: a direct and immediate injury to tissue and a delayed effect which is revealed as cancer up to twenty-five years later, or as deformed births in succeeding generations. Most of the knowledge of major damage to human beings comes from the victims of American bombs on Hiroshima and Nagasaki, and the Japanese fishermen who were caught in the fall-out from nuclear-weapons testing in the Pacific. The number of accidents in the nuclear-power industry itself has fortunately been very small. The medical statistics on people

Table 24. Types of nuclear reactor

NAME	ACRONYM	FUEL	COOLANT	MODERATOR
Magnox	—	Natural uranium	Carbon dioxide	Graphite
Advanced gas-cooled	AGR	Enriched uranium	Carbon dioxide	Graphite
High temperature gas-cooled	HTGR	Enriched uranium	Helium	Graphite
Boiling water reactor*	BWR	Enriched uranium	Boiling light water	Light water
Pressurized water reactor*	PWR	Enriched uranium	Pressurized light water	Light water
Canadian deuterium uranium	CANDU	Natural uranium	Pressurized heavy water	Heavy water
Liquid-metal fast breeder reactor	LMFBR	Plutonium dioxide, uranium dioxide	Molten sodium or molten sodium and potassium	None required

*Both these reactor types are also referred to as light-water reactors with the acronym LWR.

exposed to lower levels of radiation in uranium mining and at work as radiographers and X-ray operators, however, confirms the general conclusion that every exposure to radiation causes some damage.

A particularly insidious danger is caused by the fact that stable and radioactive isotopes of the same material are chemically identical. Nature concentrates some elements within parts of the body – iodine in the thyroid for instance; it also successively concentrates substances along food chains. Even if a radioactive isotope of one of these elements is dispersed sufficiently widely to be biologically almost completely innocuous, natural processes can concentrate it dangerously. An accident occurred at the United

Kingdom reactor at Windscale in 1957 in which a considerable amount of radioactive material was released into the atmosphere. One of these substances was iodine-131. It was found that this became concentrated alarmingly in cow's milk. All deliveries of milk from the area had to be stopped until the radioactivity had decayed to a level which was considered acceptable.

The radioactive isotope is therefore akin to a booby trap. It is indistinguishable from the stable isotope of the same material. It is used by nature as though it were stable. It can make its way into living creatures and reach damaging concentrations within vital organs; it can even be incorporated within the complex molecules of the genetic code.

The genetic damage caused by radiation is one of the most worrying aspects of the use of nuclear power. The test explosions carried out at Bikini Atoll in 1946 have left a grim demonstration of the long-term effects of radiation in the legacy of bizarrely deformed plants and animals on the island. Increasing the amount of radiation in the world inevitably increases the number of genetic mutations and consequent deformed human births.

Levels of exposure to radiation are strictly controlled in all countries working with nuclear power. The amount of exposure of workers and the public is kept to a permissible or 'safe' proportion of the background radiation caused by naturally occurring radioactive materials and the small amount of radiation from space which penetrates the earth's atmosphere. This natural radiation is itself damaging; people in areas of higher-than-average background radiation have a higher incidence of cancer and there are fears about the health of the crews of high-flying supersonic aircraft. There is, in fact, no such thing as a truly 'safe' dose of radiation and it is disingenuous to use the term to imply that no damage occurs. The nuclear-safety debate is about *levels* of damage and the question of whether some risks are permissible at all.

The most popular fear of nuclear power stations, however, is completely unjustified. Under no circumstances could a reactor containing natural or slightly enriched uranium fuel explode like an atomic bomb. In the case of the breeder reactor, however, this remote possibility does exist. If there were a catastrophic melting of the core the highly enriched fuel could, theoretically, achieve the configuration necessary for an albeit inefficient, but very definite, nuclear explosion.

The worst accident which designers consider, the 'maximum credible accident', is a complete failure of the coolant supply. If this happened, even if the reactor shut-down mechanisms operated perfectly, the heat emitted by the accumulated fission products would be sufficient to cause the whole core assembly to melt. The possibility that the pool of molten material would then burn downwards through the floor of the building has, in America, the mordantly apt name of the 'China syndrome' because of the general direction of progress of the molten mass.

To prevent this, emergency core-cooling systems are installed. A difficulty with these is knowing whether they will work in the extreme situation in which they would be required. They can only be fully tested by wrecking a series of nuclear power stations, and there is an understandable reluctance to do this. Since, however, a melt-down of the core is an accepted possibility and its consequences, if uncontrolled, would be truly disastrous – a massive release of radioactive material, in addition to causing death and disease, could necessitate the permanent abandonment of the area affected – it is essential that emergency cooling systems can be demonstrated to work. There is a furious and highly technical debate on this subject in the United States, where, in comparison with other countries, nuclear matters are discussed with commendable openness. Elaborate computer simulations of reactors under various conditions are used to predict how the emergency cooling system would behave. But it is a weirdly unreal world, divorced from the ingeniousness of human fallibility which finds ways of making mistakes beyond the dreams of most technologists.*

*It is unlikely any computer simulation envisaged the possibility of a man with a candle setting fire to cables under the control room, thereby disabling the emergency core-cooling system. But this actually happened in March 1975

It is a disturbing thought that these simulations are the only guide to the behaviour of the last-ditch defence against a major accident in nuclear power stations now in operation or under construction.

Nuclear engineering is at the edge of technical knowledge. Even its most fervent proponents admit that it is dealing with processes which are uniquely dangerous. Perhaps the best metaphor is that used by Alvin Weinberg, one of the early leaders of the US nuclear effort, when he called the use of nuclear power a 'Faustian bargain' – a promise of inexhaustible energy but one that could bring disaster if uncontrolled. It is true that the safety record of the industry has been extremely good in comparison with most other large-scale industries, and that much of the criticism voiced against it has been ill informed and hysterical. But the need for caution and humility is great.

The failure of steel bridges and welded tankers after the Second World War was because of the, up to then, unknown phenomenon of rapid crack propagation which can occur in stressed steel when the temperature is low. The tragic crashes of the first Comet aircraft were a result of unforeseen metal fatigue failure. The collapse of the Tacoma Narrows suspension bridge occurred because no one had thought structural resonance could be induced by a wind of moderate speed. The more recent rash of failures of welded box-girder bridges in the UK, Europe and Australia was because designers had not fully understood the behaviour of steel diaphragms subject to heavy buckling loads. Engineering has a long history of finding things out the hard way while at the same time making bland statements about how safe and well understood everything is. There is ample historical justification for an uneasy feeling that there is not yet a sufficient background of experimental and operating experience to justify the confidence of the nuclear

---

at the huge Brown's Ferry nuclear station in Alabama. It took seven hours to bring the resulting fire under control. Luckily, it proved possible to use other pumps, not intended for use as emergency systems, to prevent the melt-down of the reactor core.

power industry in its own ability to anticipate and deal safely with any problems which may arise from long-term reactor operation.

In addition to guaranteeing the safety of reactors, the nuclear industry must also ensure that no damage can result from possible accidents in the fuel-processing and waste-disposal operations. Some of its most intractably difficult problems are found here.

Contrary to what might be expected, the fuel going into a nuclear power station is a minor hazard, requiring little shielding to make it completely safe to handle. Natural-uranium fuel rods can be packed in cardboard boxes. But once it has been inside the reactor the fuel becomes a very nasty and dangerous substance indeed. Used fuel contains fission products, which are usually highly radioactive, and plutonium created from the uranium-238. One of the reasons fuel elements are always clad in some strong material is to contain these radioactive substances and prevent them leaking into the coolant system which would then carry them into the heat-exchangers or electricity generators. The fuel cladding material, because it is subjected to the intense neutron flux which occurs in the reactor core, itself becomes radioactive.

When the spent, or poisoned, fuel elements are removed from a reactor they are dumped in a deep cooling pond until the shortest-lived isotopes have decayed sufficiently to allow the fuel to be transported to a reprocessing plant. It is taken there in reinforced casks which must be cooled in transit. In the reprocessing plant the uranium and plutonium are separated and stored for re-use. The remaining materials must be isolated until their radioactivity has decayed to a level low enough for them to be released into the biosphere. Some authorities consider this is reached after about 20 half-lives. In the case of the common fission products, strontium-90, with a half-life of 28 years, and caesium-137, with a half-life of 30 years, the isolation period required is therefore about 600 years. Some of these high-level wastes, as they are called, are highly corrosive and, being radioactive, they all emit heat. Indeed it has been suggested that for the first five years of storage they could be used as a source of heat energy at a temperature of about 400°C.[25]

Whatever is done about this possibility, they must be sealed into tanks which are resistant to corrosion and equipped with non-fail refrigeration systems. Wastes now going into storage will not become 'safe' until the year 2575, a time as far into the future as Chaucer's is into the past. Plutonium, with its very long half-life, is for practical purposes indefinitely dangerous. In this it resembles many of the other noxious substances which industrial society discharges into the environment. Most of these too will last for ever.

The nuclear industry's present aim is to devise a method of solidifying high-level wastes, most probably by vitrification. This would make the wastes much more manageable and greatly reduce their volume – about 4 cubic metres of treated and solidified waste would be produced by a 1 000 MW power station per year. The blocks of vitrified wastes would be stored for perhaps thirty years in underground caverns during the period of most vigorous radioactivity and heat emission. After that they would be buried deep in the earth – perhaps a kilometre down – and sealed off. Alternatively, they could be buried in the sediments of the deep oceans. Although these repositories would then need neither surveillance nor maintenance, they might be vulnerable to natural phenomena. The possible disturbances caused by the next Ice Age, for example, have been given serious consideration.

Vitrification processes are now in an advanced stage of development in France; programmes are also under way in the US, the USSR, the UK and other countries. A satisfactory process should be perfected and in operation within the next decade. In the meantime the processing and short-term storage of wastes is causing multitudinous problems. Leaks of high-level waste have been regularly reported in the US and the UK. Reprocessing plants have proved to be technical and economic disasters, particularly in the US. At present no commercial-sized reprocessing plant is in operation, except in France where the plant has been the object of vociferous criticism by the trade unions for what they feel is an unacceptably low standard of safety.

In the absence of reliable reprocessing facilities, the question of vitrification and long-term disposal of high-level wastes is academic. For the moment these wastes remain in the used fuel rods – a growing and embarrassing monument to the over-confidence of the nuclear industry on this subject. In some parts of the US it is now feared that the lack of suitable storage space for used fuel may even lead to the closure of some nuclear power stations.

There are other waste products from nuclear power stations and fuel-processing plants. These are large industrial establishments in which machinery is constantly being serviced, repaired, and renewed. Anything which is used in radioactive areas becomes contaminated. Ancillary reactor parts, valves, pipes, switches, rags, protective clothing, the whole of the industrial detritus has to be carefully collected and buried at the site, or put in casks and dumped elsewhere.

Low-level wastes, such as air which has been used in the reactor ventilation system or water from the generator cooling system, are simply dispersed. High chimney stacks take radioactive gases like krypton-85, which has a half-life of 4½ hours, up to levels which allow them to be widely dispersed before they reach the ground. The mildly radioactive cooling water is discharged several miles off-shore. Continuous monitoring must be carried out to check that dangerous concentrations of radioactivity are not building up in the environment or being concentrated along food chains. Bass, for instance, produce a concentration of caesium in their flesh which is 1 000 times that in the surrounding water; while plankton have been recorded as concentrating strontium-90 up to 75 000 times the level in the water in which they live.

The biggest waste product of all is the reactor itself. Once it has been in operation, a nuclear reactor becomes highly radioactive. The materials of which it is constructed are subjected to neutron bombardment and many of them turn into radioactive isotopes with long half-lives. When a reactor has reached the end of its useful life through mechanical breakdown, wearing out, or corrosion of materials, there is very little to be done except leave it, sealed against human entry, for a couple of hundred years. Theoretically it might be possible to dismantle it by remote control but in practice this is hardly likely.

With the passing of time it is becoming increasingly difficult to form a confident opinion about the future path of nuclear development. Well-informed appreciation of the dangers inherent in nuclear power has been growing and no one now denies that they are uniquely disturbing ones. Its supporters maintain that the risks can be kept to an acceptable level by dedicated care and attention. But accidents such as the spectacular incident at Three Mile Island have, not surprisingly, undermined public confidence in the industry. Air crashes, computer breakdowns, and the losses and failures of space satellites, are recurrent evidence of the fallibility of advanced technology and the people who operate it. The fact is that the nuclear industry must aspire to tighter standards than conventional industry. An accident in a coal- or oil-fired power station which would scarcely make the front page of a local newspaper can, in the case of nuclear power, attain the status of a national emergency.

Also disturbing is the threat of nuclear wars, and the possibility of terrorist blackmail. The enriched uranium necessary for a bomb is beyond the capacity of any but the major nuclear powers to produce. But separating plutonium from the irradiated fuel rods of a nuclear reactor is, in comparison, a minor operation, and well within the reach of a large number of countries. Several of them are undoubtedly working on such projects. Nuclear weapons can easily be developed secretly, under the cover of a nuclear power programme. Meanwhile fuel-processing plants, nuclear reactors, waste maintenance sites, and the radioactive transports between them could make inviting targets for political or criminal fanatics.

The anti-nuclear polemic is no longer exclusively based on the physical dangers of exploiting atomic energy. Opponents of nuclear power have voiced their apprehension that civil liberties might be reduced, or democracy eroded, to safeguard the 'nuclear society'. Furthermore, they charge that the nuclear industry and national governments make their decisions behind closed doors and withhold vital relevant information from the public. The feeling among many who protest against nuclear power is that society is being secretly and irrevocably committed to a dangerous and repressive society under the guise of providing for future energy needs.

In this climate of mistrust and implacable opposition level-headed argument is usually to no avail. It can be demonstrated that coal-fired electricity generation is many times more dangerous than nuclear – some coal-fired stations, in fact, discharge more radioactivity than a nuclear station. But there is little sign of a passionate anti-coal movement. The symbolic appeal of the anti-nuclear case, whatever its real basis, is very strong and likely to endure.

Nuclear power programmes are also plagued by construction delays, recurrent technical problems and escalating costs, which distract attention from their potential competitiveness. The capital cost of nuclear power stations, with waste storage and fuel-processing facilities, is considerably higher than that of conventional coal- or oil-fired stations. But nuclear power has a major advantage in its lack of sensitivity to changes in the price of its fuel. The cost of uranium is such a small proportion of total generating costs that increases in its price, even on the scale of those for oil in recent years, would have comparatively little effect on the price of the electricity produced. Although the figures are widely disputed, there is growing acceptance that nuclear power is competitive with oil-fired electricity generation, and could even be competitive with coal.

Paradoxically, the 'energy crisis' of the 1970s, which was seen by many as a signal for accelerated development of nuclear power, has been extremely damaging to its prospects. This is because the demand for nuclear power depends specifically on increased demand for electricity; but electricity consumption tends to increase only in an affluent and expanding economy. A great deal of it is used to provide improved standards of living, comfort and even luxuries. The economic recessions of the 1970s have slowed or halted the demand for more electricity in most industrialized countries. As a result, the demand for new power stations has been much smaller than was anticipated.

Since the heyday of its early commercial development in the 1960s, when it was confidently expected to resolve all the world's energy problems, proponents of nuclear power have constantly revised their forecasts of its future role downwards. In a comprehensive study published as late as 1974 the OECD estimated total installed nuclear generating capacity for the world in the year 2000 at 2 800 gigawatts* and reckoned that an accelerated programme could provide 4 100 gigawatts. The most recent estimate by the International Atomic Energy Agency is that the year 2000 will see a nuclear capacity in the range 1 000–1 600 gigawatts.

Existing installed world capacity is about 120 gigawatts. Completion of plants under construction or on order would bring this to just over 400 gigawatts by the late 1980s. This could well be regarded as the maximum credible contribution from nuclear power by 1990. The flow of orders to the industry has slowed to less than 10 gigawatts per year, a fifth that of the early 1970s. Moreover, it is far from certain that all the stations on order will be built. This is particularly true in the US where many planned stations are being delayed by regulatory proceedings, and electric utilities are seriously reconsidering their previous nuclear plans. Coal is emerging as an economically attractive and far less contentious competitor in a considerable number of cases.

Figure 10 shows the way OECD estimates of installed capacity by the year 1985 have fallen over the past ten years. It accurately portrays the gathering disenchantment with nuclear power in the 1970s. Unless there is an unforeseen rapid increase in orders over the next few years, the world's total installed capacity is unlikely to be more than 500–600 gigawatts by the year 2000. Although far short of what the nuclear industry would like to see, this would still provide a large amount of energy. It is equivalent to the output of coal-fired power stations consuming about 1 billion tonnes of coal a year, about 10 per cent of the world's present total energy consumption.

*A gigawatt is one million kilowatts.

Figure 10. Past projections of OECD nuclear generating capacity in 1985

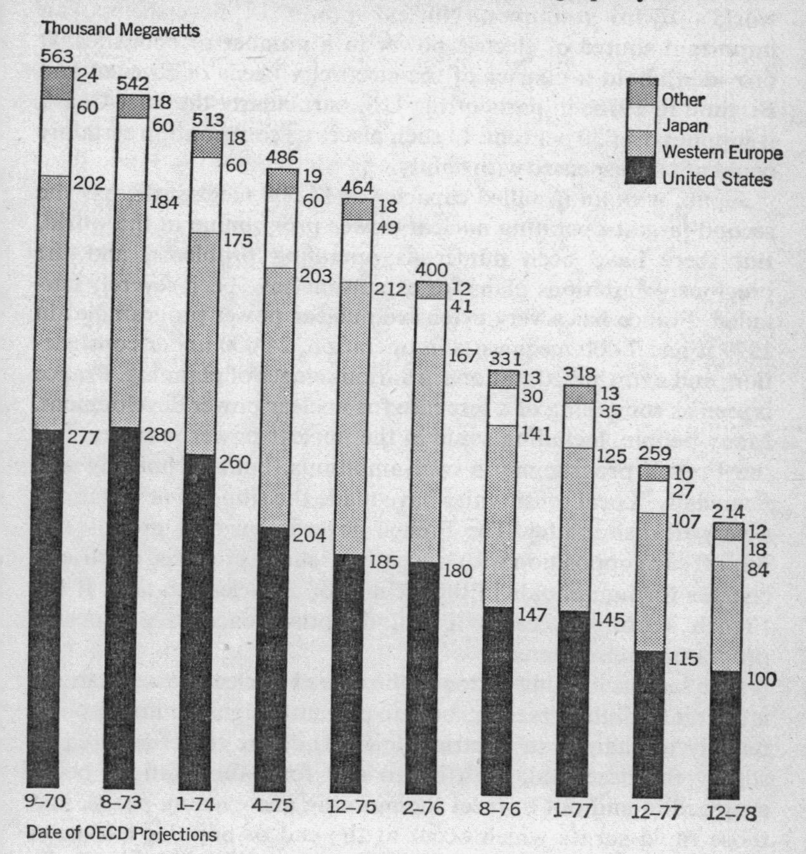

Source: *The World Oil Market in the Years Ahead*, National Foreign Assessment Centre, CIA, August 1979.

At present nuclear power provides about as much energy as would be supplied by burning 250 million tonnes of coal; that is about 2¼ per cent of the world's primary energy. It is therefore far from being a world resource on the scale of the fossil fuels. Indeed, it provides little more than a third of the energy supplied by the world's hydro resources. Nuclear power is, nevertheless, an important source of electric power in a number of countries. It provided about a quarter of the electricity needs of Sweden and Belgium in 1978; in parts of the US, particularly the East Coast, it supplies over 30 per cent. In such places its contribution certainly could not be dispensed with lightly.

Japan, with an installed capacity of 14 500 megawatts, has the second-largest operating nuclear power programme in the world. But there have been numerous operating problems, and the previously ambitious plans for expansion have been severely curtailed. France has a very extensive nuclear power programme: in 1979 it had 7 400 megawatts in operation, 24 000 under construction, and a further 20 stations in various stages of planning. France is seen as something of a test case for nuclear power development. Many people, including some in the nuclear power industry, feel the French programme is over-ambitious, both technically and financially. Local sensitivities have made the siting of new stations a very ticklish matter. The French government now proposes to buy off such opposition with 'incentives' such as reduced electricity charges for households in the vicinity of a nuclear station. If the French venture succeeds it will doubtless encourage renewed programmes elsewhere.

One serious limiting factor to the role of nuclear power stations is their inflexibility; reactors built to present designs cannot respond quickly to changes in electrical load. They are therefore used to supply the base-load, with hydro and fossil-fuel stations being switched in and out to meet the morning and evening peaks, and those rapid surges which occur at the end of popular television programmes. The closer nuclear power comes to meeting the base-load the more difficult it becomes to accommodate further nuclear stations within the generating system because no simple method of storing large quantities of electricity has been devised. All forecasts which envisage nuclear power supplying a high proportion of a country's electricity consumption are implicitly based on new reactor types with good load-following characteristics. Such reactors do not yet exist.

On the plus side, fuel supply is unlikely to become a limiting factor. Uranium is, in fact, a relatively common element. It is found in low concentrations in granite and in sea water. In a more concentrated form it is found in the ore pitchblende, and as a constituent of many shales. The Colorado shale contains some, and it is found in richer concentrations in the Chattanooga shale which underlies much of the states of Tennessee, Kentucky, Indiana and Illinois. It is also found in various conglomerate rocks in sedimentary deposits. According to the Canadian Department of Energy, Mines and Resources hundreds of thousands of square miles of the country are 'geologically favourable for uranium, and many are relatively unexplored'.[26]

Although it was discovered in 1789, and was mined, first as a scientific curiosity, and then as a source of radium, uranium has been regarded as an important resource only in the past few decades. It has had no long history of exploration, as have had the common metals, coal, and oil. Large-scale mining of uranium only began, in Canada, in 1942 in response to the military programme in the United States and Britain. Once these requirements had been fulfilled there was an almost total collapse in the market and 'uranium exploration in Canada was essentially dormant from 1956 to 1966'.[26] A short burst of activity occurred between 1966 and 1969 but exploration again declined. Since 1974 exploration has intensified and is disclosing plentiful recoverable resources. Accessible reserves could also be greatly increased by new mining techniques or higher prices, which would make the mining of lower grades profitable.

Table 25 shows the countries with major uranium deposits. Resources recoverable at up to $26 per kilogram (1974 prices) amount to just under a million tonnes. But increasing the price by only 50 per cent, to $39 per kilogram, quadruples the total. The 1974 World Energy Conference *Survey of Energy Resources* comments: 'At costs up to $200 per kilogram the amounts of

Table 25. Countries with major uranium resources

Country	Reasonably assured resources up to $26/kg uranium	Total uranium* resources
	tonnes	
United States	329 267	2 041 156
Canada	185 799	716 984
Sweden	—	308 381
South Africa	202 000	298 004
Australia	120 949	160 049
France	34 850	85 000
Nigeria	40 000	80 800
India	—	61 862†
Colombia	—	51 000
Argentina	12 665	38 590
Gabon	20 400	30 240
Rest of Europe	21 834	73 863
Rest of world	16 710	78 023
TOTAL	984 474	4 023 948

Source: World Energy Conference, *Survey of Energy Resources* 1974.

*Includes 'reasonably assured' and 'estimated additional' resources at costs of up to $38 or $39/kg uranium with small amounts at unspecified or higher extraction costs.

†At costs up to $77/kg uranium.

uranium available are in tens to hundreds of megatonnes* and at costs up to $500 per kilogram in thousands of megatonnes'.[27] This is not to say that there would be no problems in mining ores of these low concentrations. The mining itself would consume large areas of land and there would be gigantic amounts of waste. Recovering a tonne of uranium from an ore in which it occurred at a concentration of 0·005 per cent, typical of the Chattanooga shale, for example, would leave 20 000 tonnes of waste. Mining and upgrading such ores would also be very expensive in energy.

The present production of uranium is about 23 000 tonnes per

*A megatonne is a million tonnes.

year, excluding the Communist countries. 'Reasonably assured' reserves could accommodate this consumption for nearly fifty years, while moving to reserves costing $39 per kilogram would provide enough for 200 years. Consumption is slowly increasing, but at nothing like the rate envisaged in some former projections on which forecasts of early depletion of usable uranium resources depend. Without being complacent, it seems reasonable to assume that world uranium resources are adequate to supply any likely nuclear development programme for at least the next hundred years. There is also considerable room for improvement in fuel utilization in existing types of station, which would lessen the pressure on resources.

It follows that scarcity of uranium cannot be a pretext for a crash programme of breeder reactors. This has been promoted as the only way to counteract a critical depletion of uranium reserves but any such crisis now looks a long way off. While a successful breeder reactor would undoubtedly use uranium much more efficiently there are many reasons for proceeding slowly and cautiously – and none for rushing.

Frederick Soddy, with remarkable prescience, said in 1912: 'When coal is exhausted and the other physical resources of the earth have been squandered, when expanding civilization is met by a dwindling supply of energy, either science or the atom will have been tested to destruction and one or the other will be the arbiter of the future.'[16] Nothing has happened since then to change this view. The energy within the atom is still the only source man has identified which is capable of supplying the needs of industrial society after oil and coal are gone. But there are no grounds for complacency: the atom has not yet been satisfactorily tamed. With existing technology, it cannot substitute for fossil fuels in all their varied uses. Lastly, but not the least important of reservations, nuclear power is still almost totally irrelevant to the development needs of the Third World.

Fusion Power

Nuclear fusion is the source of the energy of the sun and the stars. In the extremely hot cores of these bodies atoms of the lightest elements coalesce to form heavier elements. Some mass is lost in the process and released as energy. If the same process were harnessed on earth in a fusion reactor, many people believe humanity's energy problems would be solved for ever.

Fusion reactors, however, do not exist. They are products of the technological imagination. The nearest man has come to harnessing fusion power is in the hydrogen bomb. It remains to be seen whether it will ever be possible to control a fusion reaction and make use of its energy. Professor Lidsky of the Massachusetts Institute of Technology has summed up the attitude of probably the majority of scientists working on nuclear fusion: 'The search for economically controlled fusion power is a scientific hunt for the Lost Dutchman Mine. Only a few believers are *absolutely* certain that the goal exists but the search takes place over interesting ground and the rewards for success are overwhelming.'[28]

About thirty different energy-releasing fusion reactions are theoretically possible but only those using the hydrogen isotopes deuterium and tritium* and the common metal lithium are of practical interest. If all the deuterium in 1·5 cubic kilometres of sea-water were used in a fusion reactor the energy released could be equal to that in the world's recoverable crude oil. Since there are about one and a half billion cubic kilometres of sea-water the rewards for the successful development of fusion power are indeed overwhelming.

There is a simple and engaging logic about the processes which would occur in a fusion reactor. When two deuterium atoms combine they either form a helium atom and release energy, or they form one tritium atom and one ordinary hydrogen atom, also releasing energy. The tritium in turn combines with deuterium,

*Tritium is hydrogen with two neutrons in its nucleus and hence an Atomic Weight of 3.

again forming an isotope of helium and releasing energy. If lithium is introduced it is broken down into helium and tritium (a process of nuclear fission) with a release of energy and the tritium fuses with deuterium. All the radioactive components are thus 'consumed' and the final products of the whole reaction are helium, hydrogen and energy. As, however, there is a neutron release during the reaction, the containment vessel and exposed equipment would become radioactive, though probably not to the same extent as in a fission reactor.

Its logical simplicity, however, does not mean the process is easy to achieve in practice. Fusion can only occur between the nuclei of atoms which are brought into contact with sufficient kinetic energy to overcome the extremely high forces with which they repel each other. Such conditions can be obtained at a temperature of about 100 million K. At this temperature substances are not only gaseous, but they exist in a state in which their atoms have been stripped of their electrons. This high energy condition of dissociated nuclei and free electrons is called a plasma.

Obviously such a plasma cannot be kept in a material container. It would instantly vaporize it. The only way yet envisaged of holding a fusion plasma is by the use of an immaterial container made up of very strong magnetic forces. The problems of holding the plasma stable at the correct temperature and extracting energy from it are well beyond the present powers of science. Theoretical and experimental work to date has, therefore, not even attempted to produce all the conditions necessary for an energy-producing reaction. Instead it has been concentrated on separate aspects of the problem.

Theoretical solutions are emerging, but the next stage is more difficult because experiments to confirm the theoretical results have yet to be devised and carried out. Many of these require hundreds of millions of dollars' expenditure and years of work with no guarantee of producing anything more constructive than the conclusion that a theoretically attractive approach does not work in practice. In the early 1960s there was much premature, and ill-informed, enthusiasm for a British device named ZETA which was supposed to have answered most of the questions. It had answered some but its use revealed a host of problems previously unsuspected. Extensive work on fusion research is now being carried out in Russia, the US, Japan and Europe. The basis of most of this is the Tokomak machine pioneered by the USSR. This uses a large circular ring within which the plasma is confined by means of magnetic forces.

It is difficult to make an assessment of how close researchers are to constructing a working fusion reactor. Some scientists have an uneasy feeling that the present approach may never be successful. One criterion of progress is known as the 'Lawson Product'. This is the product of the density of the plasma in ions per cubic centimetre and the confinement time. When a plasma meeting the requirement of the minimum value of the Lawson Product is confined at a temperature of 100 million K it becomes theoretically possible to extract more energy from it than is being put into heating and confining it. The minimum necessary 'Lawson Product' is $1{\cdot}0 \times 10^{14}$ seconds per cubic centimetre. Experiments to date have given results with a Lawson Product value of $1{\cdot}0 \times 10^{13}$, which is about a tenth of that required.

An alternative approach which is receiving increasing attention relies on using laser beams focused on tiny pellets of deuterium and tritium in such a way that they are compressed and heated sufficiently for the fusion reaction to occur. Work on this is being taken very seriously in the United States and funding for it is now approaching the same level as that for 'traditional' fusion research. But development is still at a very early stage and no predictions about its success can be made.

A quotation from one of the leading scientific groups working on this method of laser 'implosion' speaks eloquently for itself:

> In addition to a substantial advance in laser technology, a laser-fusion power plant will require the solution of many other technological problems. The high-efficiency detonation of fusion-fuel pellets for practical electricity generation will occur on a time scale of 10^{-11} seconds

or less. Since the energy released will be at least 10^7 joules, the peak rate of fusion power production will be at least 10^{18} watts. This rate (which, to be sure, is intermittent) is a million times greater than the power of all man-made machinery put together and is about 10 times greater than the total radiant power of sunlight falling on the entire earth. The technological challenge of laser fusion is to wrap a power plant around fusion micro-explosions of these astronomically large peak powers that can endure their effects for dozens to hundreds of times every second for many years.[29]

At this stage of uncertainty there is no point in dwelling on the amount of fuel available for fusion reactors. For practical purposes it can be regarded as infinite; sieving the oceans of their deuterium could continue for millions of years. Lithium can also be recovered from sea-water and could eventually provide a near-infinite source of supply of this element at essentially constant costs. An interesting discussion by Kulchinski,[30] however, suggests that the availability of materials for constructing fusion reactors could well be a limiting factor. Any potential fusion reactor would appear to need large quantities of comparatively rare materials such as vanadium, niobium and molybdenum.

It is obviously unreasonable to go too deeply into the question of the safety, or otherwise, of fusion power. The potential hazards seem to be fewer than those associated with fission-power stations. If it can be mastered, fusion is likely to provide electricity more cleanly than any other method.

In summary, then, fusion power is an attractive but still distant promise. Although some optimistic scientists believe that a theoretical demonstration of the feasibility of a fusion reactor might be possible by the late 1980s, such a breakthrough would be just one step in the right direction. It would have to be followed by laboratory demonstration, operational prototypes, and commercially practical design, before any programme for a network of stations. None of these steps is unproblematic. The world will surely have to wait at least fifty years for any help from fusion power.

11

'Non-Depleting' Energy Sources

The sun, the rivers, the tides and waves of the sea, the wind, the earth's internal heat, and the natural growth of plant life can all be used as energy sources. They are often classed together and referred to as 'non-depleting'. But this is not accurate. Only the sun, the wind and the waves remain unaffected by man's use and are truly non-depleting over centuries.

It is more correct to think of the others as depleting. In the case of hydro or tidal power the installations used for channelling and damming the water become silted up and eventually unusable. The river or estuary may then have to be abandoned as a source of power. Geothermal energy is only obtainable at certain favourable sites, usually reservoirs of hot water or steam, and these are depleted when the heat is drawn from them faster than its natural rate of renewal. And timber, or any other plant life used as fuel, cannot be cropped without depleting the soil of the nutrients on which renewed growth depends.

Solar Energy

The most tantalizing feature of the sun's energy is its abundance. The solar flux at the outer edge of the atmosphere is 1·4 kilowatts per square metre. Although absorption and reflection reduce this, about 50 per cent of it reaches the earth's surface. The noon intensity of solar energy on a clear day in the tropics can exceed a kilowatt per square metre. This energy falling on an area about 8 kilometres square is equivalent to the output of the whole of the United Kingdom's electricity generating system.

The desert areas of the world extend over about 20 million square kilometres. On this area of land, which grows no food, and supports no population, the total annual solar radiation is about four hundred times the world's present energy consumption of all kinds. This obvious superfluity of energy, with no known way of collecting more than a tiny fraction of it, makes solar-energy research both extraordinarily challenging and maddeningly frustrating.

The most commonly used collecting device is the flat-plate collector. In its simplest form it is no more than a flat black surface with water trickling across it. Evaporation losses can be cut down by running the water through pipes embedded in the absorbing surface or by using a device like a flat central-heating radiator. Insulation at the back and sides improves performance by cutting down heat losses to the surroundings. Convection and re-radiation losses from the absorber surface are reduced by covering it with one or two glass plates. Glass has the useful property of being almost transparent to most of the energy from the sun and almost completely opaque to the long-wave radiation emitted from an object which has been heated by the sun. The glass plate over the collector therefore absorbs heat emitted by the absorbing surface and radiates some of it back again.

A further refinement is the use of a selective absorber. This is usually a polished metal plate covered with a thin layer of a black material such as oxide or sulphide of nickel. At any particular temperature the rate of heat emission from a polished surface is less than that from a black one – a shiny teapot keeps the tea warm for longer than a similar one with a black surface. On the other hand a black surface absorbs heat more effectively than a polished one. The selective absorber is one of those happy instances in which the best of two worlds is provided: it absorbs like a black body and emits like a polished one. The black layer, however, has to be applied under tightly controlled conditions, usually by electrolytic deposition; and it is delicate and easily worn away. Its performance is also seriously impaired by dust, which can be a big problem in desert conditions.

Another application of solar energy is the distillation of water.

Solar distillation was being successfully used in Chile in the last century and since then the basic method has not changed much. Salt water flows along trays with a black heat-absorbent floor. When the sun is shining the heat absorbed causes the water to evaporate and it condenses on a sloping glass cover plate and runs off into a collecting channel. The output of a solar still is low; it produces about 3 litres per square metre a day. The capital costs of large solar distillation plants make it an unlikely choice to meet the needs of irrigation or industry. But for small-scale use a variety of designs has been developed, many using low-cost materials, and these could be useful in the developing world to provide limited supplies of drinking water.

The main use of flat-plate collectors is in the provision of domestic hot water in countries with plenty of sun. Australia, Japan, Israel and the southern States of America are good examples where collectors are commercially available. In northern countries their use will inevitably be restricted to topping up other heating systems during the sunny part of the year – at least until the discovery of a cheap and effective method of long-term storage of heat which would enable households to store summer heat for the winter.

Very large sums of money are being spent on research into exotic refinements of the flat-plate collector, some of it by companies mainly devoted to aerospace technology. While some improvements in efficiency and general thermal performance may result from this research, most of it seems far removed from the mundane requirements of domestic water heating. It is not the extra few per cent of thermodynamic efficiency which will attract the householder; the real need is for cheap, durable and simple systems which can be installed and maintained easily.

Much higher temperatures than those common in flat-plate collectors can be obtained by using focusing collectors. A 'magnifying' glass is one such device, and Lavoisier used a lens as big as a man for some of his experiments in 1774. The application has an even longer history if Archimedes really did use focusing mirrors to set the Roman fleet on fire in 212 BC.

The modern focusing collector uses a mirror, rather than a lens. The temperature at the focus is determined by the area of the collector and the precision with which the energy is focused. Simple concave collectors about 1·2 metres in diameter can bring a litre of water to the boil in about 15 minutes, but temperatures as high as 4 000°C have been achieved in the Odeillo furnace in the French Pyrenees. Here, ranks of adjustable mirrors on a hillside reflect sunlight down on to a large concave mirror, which in turn focuses on to a furnace. The absolutely clean heat supplied and the high temperature are useful in many research programmes. But the focusing collector has two major drawbacks. Whereas a flat-plate collector will pick up heat if pointed in the general direction of the sun, the focusing collector must be precisely directed at the sun if it is to operate at all. This means it requires constant manual adjustment or a sophisticated tracking mechanism to keep it in the required position. The flat-plate collector, moreover, is able to use the indirect radiation of the sun which has been scattered by clouds or atmospheric constituents, but this energy is useless to the focusing collector. Even hazy conditions put it out of action.

There were high hopes, particularly during the 1950s, that cheap focusing collectors might be introduced as cookers in the developing countries, but very little has happened. The problems encountered are often as much social as technical. Many people cannot, or do not want to eat at midday, which is when it is easiest to use the cooker. The fact that the cooker is useless in cloudy weather means that people must have an alternative method of cooking in any case, and they often prefer to rely on this all the time. The continual adjustment required may also be too irksome for people unaccustomed to mechanical gadgetry.

Flat-plate and focusing collectors deliver solar energy as heat, usually in the form of hot water, or sometimes as warm air; but for many purposes this is not the way energy is required. It may be needed as mechanical power or electricity. Tapping the energy falling on the uninhabited deserts would almost certainly necessitate turning it into electricity in order to transmit it to the places where it could be used. Both mechanical power and electricity can, in fact, be obtained from solar energy, but at a price which is usually prohibitively high.

A theoretical solar-powered engine depending on heat delivered to it by a flat-plate collector at an average temperature of, say, 65°C and operating over a temperature drop of 40°C would have a theoretical operating efficiency of only 12 per cent. In practice, losses and friction would bring this down to, at best, 2 or 3 per cent. To obtain any useful power such an engine would require very large areas of collector. A number of experimental solar pumps have been installed in the sub-Saharan region but they are elaborate and extremely expensive. Typical capital costs (1979) are $50 000 per kilowatt compared with about $150 per kilowatt for a diesel pump. The solar pump also requires considerable expenditure on maintenance.

Higher mechanical efficiencies can be obtained by using focusing collectors to make steam which is then used to drive conventional reciprocating or turbine engines. Such engines have been made – one which ran a printing press was exhibited in Paris in 1878 – but mainly as curiosities. The most successful seems to have been one built in Egypt in 1913 which developed up to 40 kilowatts and was used for pumping water from the Nile. The fact that these engines can work only when there is bright direct sunshine limits their use, though it is possible to imagine circumstances in which short bursts of operation, whenever conditions are favourable, could be turned to some useful purpose. In the early part of the present century, J. A. Harrington in New Mexico used a solar-powered engine to pump water into a high-level storage tank from which it ran down to operate a water turbine and electricity generator. In general, the solar engines built so far compare rather badly with a biological solar-energy convertor such as a horse, which requires nothing more elaborate than a field of grass for its collector and storage system.

The development of commercial-scale electricity-generating plant is the goal of a lot of solar-energy research. The United States government is spending a great deal of money and again the aerospace companies are deeply involved. Large sums are also being spent in the Middle East, particularly in Saudi Arabia.

One proposed design uses a central tower at the top of which is a solar boiler. On the ground a vast field of dirigible mirrors reflects the sun's rays on to the boiler. In Italy a 50-kilowatt experimental plant has been built, using a collector array of 270 mirrors. Designs have also been drawn up for a 100-megawatt plant in the US; this would require some 25 000 movable mirrors spread over an area of about 2 square miles. These mirrors must track the sun accurately, directed by computer; they must be resistant to wind pressure and dust abrasion; and they must be easy to keep clean.

The outlook for centralized solar-power stations is that their development will be slow and expensive. Though they may be locally important, in sunny areas, it is hard to see how they can make any significant impact on world energy needs in the next half-century.

Solar energy can be converted directly to electricity using a photoelectric or photovoltaic cell: the familiar photographer's lightmeter. This uses a carefully manufactured crystal, usually of silicon, in which a small electric current can be produced when it is hit by light of the correct energy level. These crystals demand high precision manufacture: they must be made of silicon purified to one part in a hundred million and then 'doped' with an 'impurity', such as boron, to one part in a million. Silicon solar cells have been extensively used in the space programmes to provide power for satellites. They are already used for buoys and remote signalling and recording devices. Unfortunately they are extremely expensive. They cost about $10 000 per peak kilowatt of output. Costs, however, have been falling rapidly. A twenty-fold reduction in cost, similar to that common in other areas of electronic technology, would open a huge market and a multitude of applications from small-scale domestic generation of electricity for water-heating to large-scale industrial uses. Though the day when people can drop into their local electrical shop for a square metre of solar cells may be far away they could be in widespread use by the turn of the century. This is a research area of intense interest and enormous potential.

In Israel an entirely different approach to electricity generation is now at the experimental stage. This uses the 'solar pond' developed by Tabor. The pond is about a metre deep with a black heat-absorbent bottom. The lower part is filled with concentrated brine and over this is carefully poured a layer of ordinary water; the difference in density prevents the two from mixing. The brine is heated by its contact with the heat-absorbent bottom, but because of its high density it remains trapped below the water which acts as insulation. Quite remarkably high temperatures, approaching 100°C, have been reached. The heat drawn off is used to drive a low-energy turbine also developed in Israel.

A 150-kilowatt experimental installation was opened at En Bokek in 1979. A 5 000-kilowatt station is scheduled to be in operation in 1981. And by using 400sq. km of the Dead Sea it is claimed that a capacity of 2 000 megawatts could be reached by the year 2000. Israel is fortunate in having the Dead Sea, but any expanse of shallow water is apparently suitable, the salt being added artificially if necessary. Generating costs are said to be equivalent to those of hydro-electricity.

Solar energy can also be used to cool buildings. The mechanical equipment for this is akin to that used in the domestic refrigerator, which uses the mechanical energy supplied by a compressor pump to take heat from the refrigerator cabinet and expel it to the outside air. In the solar cooling device an absorption–desorption circuit is used, employing principles which were demonstrated as long ago as 1824 by Faraday. The thermodynamics are complicated but, basically, cooling is achieved by evaporating ammonia and absorbing it in a 'carrier fluid' which is then regenerated by being passed through a solar collector. Here the ammonia is re-vaporized and then passed through a condenser where it loses its heat and is liquefied to begin the cycle again. There is an inherent logic in the use of solar energy for cooling. The greatest need for cooling occurs when the solar energy required to provide it is available. This is in contrast with solar heating applications which are most needed when there is no solar energy whatsoever. The application of solar cooling for cold storage, commercial and domestic use thus has considerable potential. But the cooling systems so far developed are more elaborate, expensive and unreliable than those powered in more conventional ways.

The 'Trombe' wall shown in Figure 11 is another way of controlling the thermal behaviour of a building. A glass sheet is fixed in front of a wall facing the sun. The wall absorbs heat and in turn heats the air in the gap between wall and glass, causing it to rise. To cool the building the bottom flap in the glass sheet is kept shut and the top one opened. The heated warm air flows upwards and draws cool air into the house from the shaded rear. To heat the building the top flap is closed and the bottom left open.

Despite the efforts made over the past thirty or forty years disappointingly little progress has been made in developing practical applications of solar energy. Professor Hottel, one of the leading authorities on the subject, has commented that during the twenty-five years when he was chairman of a committee in charge of allocating the income from a large endowment for solar-energy research, the number of research centres increased from three to thirty-five and 'the common pattern, with few exceptions, was repetition of older research and the addition of little that was new and less that was economically significant'.[31]

There are good reasons for this. Although solar energy is abundant it is diffuse and difficult to collect in any but a low-temperature form. Any conversion process which turns it into a more useful form must, therefore, operate at a very low mechanical efficiency. High yields of useful energy from the sun require heavy capital investments; the energy may be free but its collection is, so far, prohibitively expensive for all except the simplest uses.

The other overwhelming difficulty with solar energy is its variability. During the day it varies from zero at sunrise and sunset to a peak at noon. The noon peak itself varies from a minimum at the

Figure 11. 'Trombe' wall

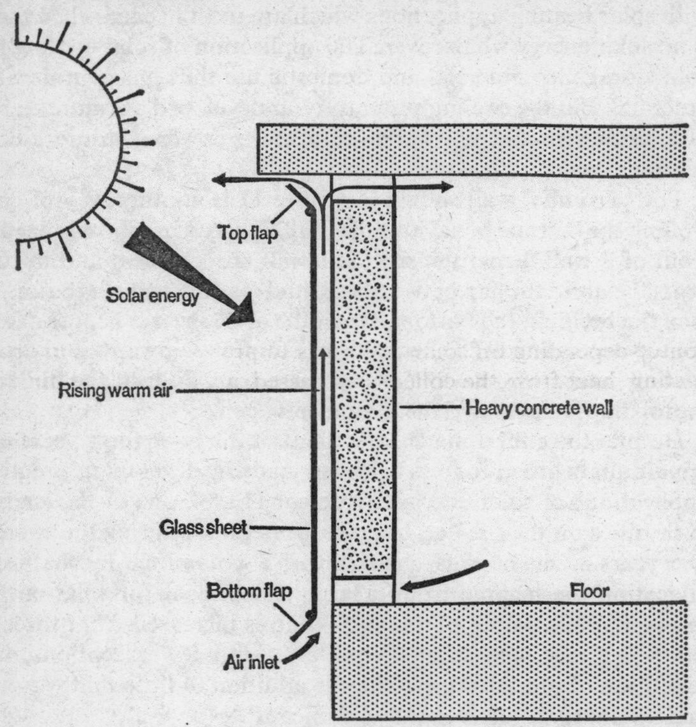

Note: The principles developed by Professor Felix Trombe and the architect Jacques Michel have been incorporated in some houses built in the French Pyrenees.

winter solstice to a maximum at the summer solstice. On top of these inherent, and regular, variations must be superimposed the random fluctuations imposed by weather conditions. The regular processes of industrial society are ill suited to such a varying and unreliable energy source. Solar energy cannot be used as the primary supply in any application where regularity and reliability are important. This means that solar-energy applications are restricted to topping up heating systems when the sun is shining, or supplying heat for swimming pools or other such purposes where interruption of supply does not particularly matter. Only in areas where there is reliable, year-round sunshine, which in fact tend to be deserts below latitudes of about 40°, can solar energy be relied upon as a primary-energy source. Many of the present difficulties could, however, be overcome if a cheap way of storing solar energy could be devised, and this is the aim of much modern research.

The simplest approach avoids any energy transformation and just relies on the capacity of materials to store heat. Thus, the water heated by a flat-plate collector during the day can be stored in a well-insulated tank and drawn off in the evening. In a 'solar house' depending on solar collectors to provide some of its space heating, heat from the collector is passed, as hot water or air, to a water storage tank or a bed of concrete or rock fragments underneath the house; in the evening the heat is drawn off and circulated through radiators or a hot-air heating system. Depending on the size of the heat store and the collector, enough heat can be retained to bridge over cold periods lasting up to perhaps a week. But this system is obviously quite expensive and almost impossible to install in anything except a large new building. An ingenious variation of this method is the use of a pool of water on the roof of the house. In winter it absorbs heat during the day and at night heavily insulated covers are drawn across and the stored heat is released downwards into the house. In summer the process is reversed: the pool is covered during the day and absorbs heat from the house, and this heat is released to the atmosphere at night when the covers are drawn back.

Such methods of solar heat-storage introduce no novel principles. They are variations on the age-old principle of using massive wall construction to store heat during the day and release it at night. At most they can store about a week's supply of energy. They are mainly applicable in areas where there is regular year-round sunshine but a sharp drop in night-time temperatures which make it necessary for houses to be provided with some form of heating system. They are of little relevance in a climate such as that of the UK where the mean daily sunshine in December and January is just about an hour and a half, and for six months of the year scarcely exceeds three and a half hours a day. Under these conditions the collector would not be able to charge the store sufficiently during the day to keep the house warm overnight, let alone carry it through a week of dark and frosty weather. To solve that problem and enable a house to obtain all its heat from the sun a method of storing heat in summer for use in winter is required.

But such long-term energy-storage is a problem on which very little progress has yet been made. Electricity can be stored in a variety of batteries but all are cumbersome and expensive. Even the automobile industry has failed to produce a cheap and effective alternative to the lead–acid battery. And the biggest car-batteries store only enough electricity to light a couple of bulbs for a few hours, as anyone who has left a car's headlights on overnight knows. To supply the domestic needs of a household for a couple of days, let alone the sunless weeks of mid-winter, would be far too expensive.

Work is also being carried out on methods of heat storage which rely on the heat absorption during a change of state from liquid to gas or from solid to liquid. When, for instance, ice melts, a large amount of heat is absorbed without any rise in temperature: it takes about 80 times as much heat to turn a quantity of ice into water as it does to raise its temperature the last degree from −1°C to 0°C. When the ice re-forms this 'latent heat' is given off. For solar heat-storage a substance is required which melts at the temperature reached by water from a flat-plate collector, and which can then be easily stored at that temperature.

The most promising substance investigated thus far is Glauber's salt ($Na^2SO^2.10H^2O$). It melts at about 32°C, dissolving in its own water of crystallization, and absorbing 67 watt-hours per kilogram – so one kilowatt-hour of energy storage would require about 15 kilograms. When the liquid is cooled slightly it recrystallizes, giving off this heat. This is, obviously, an expensive method of energy storage and to date has not been made to work particularly effectively. Repeated cycles of heating and cooling cause variations in concentration to occur in the liquid and can prevent recrystallization occurring at all within the temperature range required. The search for a better medium goes on, but with few reports of promising discoveries.

With all the attention recently drawn to solar energy it is not surprising that ideas for using the high technology of space exploration have been put forward. One proposal is that huge satellites, with arrays of silicon solar cells, should be placed in geo-stationary orbit round the earth so that they could collect solar energy and beam it back to earth as microwave radiation. While this is certainly an attractive idea for the companies involved in aerospace research, as an energy proposition it is more dubious. Hottel is bluntly sceptical:

> The proposal that ... we put up giant satellites to intercept solar energy, convert it there to microwave radiation, beam it to earth in a dilute enough form not to make a death ray of it, reconvert it to electrical energy and then transmit it to our cities appears even less attractive in its long-range possibilities. It puts in series at least four steps each one of which is far beyond our present capability except at prohibitive cost.[31]

In 1971 Hottel summarized the existing position and immediate prospects for solar-energy use as follows:

> Domestic hot water from the sun is economically significant in many areas today, solar house heating in some, and its prospects are improving; solar distillation to produce fresh water from saline water is economic in areas of extremely high fossil-fuel cost (certainly not in the US mainland); solar electric power from photovoltaic cells is significant in space research where the laws of terrestrial economics are inapplicable and it has some chance of becoming much cheaper. There are certainly enough of these areas to justify a vigorous research programme, but a major effect on the national energy picture is not to be expected.[31]

Sadly, there is still no reason to change a word of what he said. Of course there are good reasons for encouraging the use of solar energy – any easing of the pressure on fossil-fuel supplies is welcome. But solar energy will provide no easy or immediate solution to the world's energy problems.

Hydro Power

The energy of fast flowing or falling water was recognized at least as long ago as 85 BC. At about this time, a poem by the Greek poet Antipater celebrates the liberation of maidens from the task of grinding corn by hand: a waterwheel enables the water nymphs to do the work for them. Mithridates, whom Pompey defeated in 65 BC, was famous for his possession of a watermill. And Vitruvius, writing in about 20 BC, described in detail a watermill for grinding corn.

The Roman empire developed water power as a means of freeing for military use horses which were previously used to drive corn mills. It was not the first, or last, technological advance to be spurred by military objectives. Their development of water power, however, made the Romans vulnerable to interruptions in its supply. When the Goths cut the Trajan aqueduct in 536 AD an alternative source of power had to be found and Belisarius is said to have developed the floating watermill. This operated between two boats moored on the Tiber, being driven by the water flowing between them. Mills of this type were used for centuries afterwards and in medieval towns were frequently moored beneath the arches of the larger bridges. Although the output of power was small, the mill had the advantage of being able to work independently of the water level.

The Domesday Book records no less than 5 624 watermills in Britain, indicating the heavy dependence of the medieval economy on water power. At the peak of their use in Britain there were as many as 20 000 mills in action. They were used not only for grinding corn but for operating bellows and hammers in forging iron, for sharpening tools and weapons, for textile manufacture, for tanning and even for pumping water from mines. By modern standards, however, the output of power from the watermill was small. The larger mills rarely delivered more than about ten kilowatts, while the smaller domestic ones produced little more than half or three-quarters of a kilowatt.

There were numerous designs of watermill. Water was delivered through a 'flume' or channel to the top of the 'overshot' wheel, and at mid-height in the 'breast wheel' design. The 'undershot' wheel was the most inefficient and basically just dipped into the flowing stream; it only managed to capture about 20 per cent of the energy in the water. The other types, at their best, managed about three times this efficiency. All suffered from the major disadvantage of being unable to make effective use of a head of water much greater than their own diameter. John Smeaton, builder of the Eddystone Lighthouse, was one of the later designers who brought the waterwheel to its most refined form. A mill built by him at Merthyr Tydfil in 1800 was capable of producing about 40 kilowatts. One of the largest mills ever built was in the Isle of Man and delivered 150 kilowatts from a wheel 22 metres in diameter.

The waterwheel, in any of its traditional forms, was, however, a technical dead end. Further advances in harnessing water power did not come until the development of the turbine. Much of the pioneering work was done in France during the nineteenth century when Poncelet, Burdin and Clapeyron were the most prominent workers. Other notable work was done by J. B. Francis in America and James Thomson in England. Two main types of turbine were developed: the impulse and the reaction. The impulse turbine has a wheel with a set of buckets fixed around the rim. A high-speed jet of water shoots from a nozzle into the buckets, causing the wheel to spin rapidly. The most commonly used turbine of this kind is now the Pelton wheel, patented in 1880 by L. A. Pelton in California. These are usually used where there is a high head and a low volume of water flow, conditions typical of high mountainous country. Pelton wheels delivering 60 000 kilowatts have been built and, under ideal operating conditions, efficiencies of over 90 per cent are obtained.

Reaction turbines have vaned wheels inside a curved pressure casing through which the water flows. There are numerous designs which have been developed to suit different conditions. In tidal flows, or slowly moving rivers, efficiencies are low, but with a moderate or high head of water they can exceed 90 per cent. Reaction turbines with very high output capacities can be built. On the Yenisey River in Siberia there is a power station with twelve turbines each of 500 megawatts, and a further plant of twelve 550-megawatt turbines under construction. Churchill Falls in Canada has eleven 480-megawatt turbines. Units producing up to 1 500 megawatts are theoretically possible. One of the main reasons for not building such large units is the big change in load which occurs when they are switched in or out of service.

Hydro power installations are now invariably used to produce electricity. This makes good energetic sense. The use of water power to generate electricity is not subject to the thermodynamically inevitable 'heat-engine' energy losses which occur in a fossil-fuel-powered generating station. In a hydro power station up to 90 per cent of the potential energy is converted to electricity. Even for small-scale applications, therefore, it is, with rare exceptions, more efficient to produce electricity and use this to power any machinery than to use the mechanical power of the turbine directly. The old picturesque watermill transmitting the energy it took from the water through a series of creaking gears and belts was extremely inefficient.

There is plenty of energy to be obtained from the world's rivers. The theoretical amount obtainable from any river system is given by the volume of water multiplied by the drop through which it falls, but, of course, much of this cannot be recovered in practice. Dams cannot be built everywhere there is a harnessable fall in a river. Confinement in reservoirs increases the evaporation losses, hence reducing the river flow. Seasonal peaks or floods tend to exceed the generating capacity of the plant installed and have to be run to waste once the storage reservoir has been filled; during the dry season the flow may be too small for use. The potential output of a river system is usually calculated for 'average flow conditions'. It can be quite a small fraction of the figure obtained from a purely theoretical calculation of the total energy available.

The yield of every hydro-electric installation is progressively reduced by silting of its reservoirs. This is an inevitable process. All rivers carry some silt which is deposited in the still waters of the reservoir thus decreasing its storage capacity. In some locations provision can be made for flushing out part of the reservoir but this is often impossible. The wide expanses of water behind most hydro-electric dams are thus destined to become the swamps and mudflats of a later age. Conditions vary so much between river systems that no general figures can be given but the life of many reservoirs is probably not much more than a century. While a river is non-depleting, each hydro-electric station has a finite life determined by the rate at which its reservoir is filled with silt. Some rivers are so silt laden that it would be pointless even to try to harness them.

But, even with these reservations, the hydro-electric potential of the world is very large. Some individual rivers are particularly notable. The potential of the Yenisey–Angara in the USSR is estimated at 64 000 megawatts, about the same as the whole electricity generating capacity of all kinds in the UK; that of the Inga River in the Congo at 25 000 megawatts; and that of the Bramaputra in India at 20 000 megawatts.

Table 26 shows the distribution of the world's potential hydro-electric resources. It lists the possible generating capacity which each country might install, and it gives an estimate of the output which might be obtained from this. Note how much the output per megawatt of installed capacity varies. This is because of the differing flow characteristics of each river system. It is obvious from the table that some countries get very much better value than others for each unit of installed capacity. It is also noticeable that South America, Africa and South East Asia, regions poor in coal and oil, have major water-power resources, only about 20 per cent of which

Table 26. World hydro-electric potential: 'average flow conditions'

Country	Theoretical capacity usable megawatts (MW)	Theoretical annual output gigawatt-hours (GWh)
China	330 000	1 320 000
USSR	269 000	1 095 000
USA	186 700	701 500
Zaire	132 000	660 000
Canada	94 500	535 200
Brazil	90 200	519 300
Malagasy	64 000	320 000
Colombia	50 000	300 000
India	70 000	280 000
Burma	75 000	225 000
N. Vietnam and Laos	48 000	192 000
Argentina	48 100	191 000
Indonesia	30 000	150 000
Japan	49 600	130 000
Ecuador	21 000	126 000
Papua-New Guinea	17 800	121 700
Norway	29 600	121 000
Cameroon	23 000	114 800
Peru	12 500	109 200
Pakistan	20 000	105 000
Sweden	20 100	100 300
Mexico	20 300	99 400
Venezuela	11 600	98 000
Chile	15 800	88 600
Gabon	17 500	87 600
All other nations	514 800	2 011 800
WORLD TOTAL	2 261 100	9 802 400

Source: World Energy Conference, *Survey of Energy Resources*, 1974.

are yet developed. In Europe some 80 per cent of the total potential is already exploited; in America about 60 per cent.

The present installed hydro-electric capacity in the world is about 372 000 megawatts, about 16 per cent of its potential. These installations contribute the equivalent of 600 million tonnes of coal, or about 6 per cent of the world's present primary-energy consumption. Output of hydro-electricity has been growing at a rate of about 3 per cent per year.

A full development of world hydro resources is unlikely for several reasons. Water is needed for many other purposes besides electricity generation: irrigation is one of the most important and it is often impossible to combine the two uses efficiently. Also, the use of rivers for navigations and fisheries cannot always be reconciled with damming for power generation.

Hydro-electric schemes flood large areas of land. The Cabora Bassa scheme in Mozambique has a lake 250 kilometres long, covering an area of 2 700 square kilometres, with an output-capacity of just 2 000 megawatts – though there are plans to double this. As long as the land is in barren mountainous regions its loss may be acceptable, but many countries would not find it easy to flood their fertile lowland valleys, no matter how great their need for electricity. The creation of large lakes has been known to cause widespread economic and ecological damage.

Finance is also a problem: hydro-electric installations require heavy capital investments and the long construction time usually means that benefits are delayed for at least a decade. Many of the poorer countries find that they have more immediate and pressing demands on their available capital. The major powers have shown themselves willing to help finance such schemes as Aswan, Kariba and Cabora Bassa in the past but whether they will continue to do so in the world's continually changing political and economic circumstances is a moot point.

Perhaps the most important consideration of all, however, is that electricity is irrelevant to the majority of the problems with which the poorer countries are at present afflicted. A high level of electricity consumption presupposes a level of wealth sufficient to enable people to purchase electrical appliances. In the UK for instance 38 per cent of the total output of electricity is used for domestic consumers: home use is actually greater than the total amount used by industry. The peasant economy has no way of consuming electricity on this scale.

The trap in which many of the poor countries find themselves is that an increase in electricity consumption would mean an increase in all kinds of expensive imports. This is precisely what they are unable to afford. There is a danger that hydro-electric schemes in such countries, unless carefully integrated with the development of productive uses for electricity, will exacerbate rather than ease the problems of poverty, malnutrition and scarcity of capital from which they chronically suffer. The only beneficiaries of indiscriminate hydro-electric development can, too easily, turn out to be those who live in the wealthy areas of the big cities and those involved in big industrial enterprises such as aluminium smelters, which are usually outposts of the industry of the developed countries.

Although hydro power is incapable of providing energy on the scale of present fossil-fuel consumption, its development is likely to continue. It can be a very important resource in a country with steeply falling rivers and the capital and technical resources to harness them. Norway, for instance, supplies all its electricity from hydro power. The major potential for development, however, is in the underdeveloped countries. These, unfortunately, have many financial and social obstacles to overcome before they can gain access to their hydro wealth and its development is inevitably going to be slow. In the developed countries, Russia and Canada apart, a high proportion of the available resources has already been tapped and harnessing the remainder would not greatly affect the total energy-supply position.

Tidal and Wave Power

Watching the waves breaking on the shore, or the daily rise and fall of the tides, gives the feeling that there should be a simple way of harnessing this obvious source of energy. The frustrations of practical research in solar energy and in tidal or wave power have many similarities. The energy is there, and in quantity, but extremely difficult to tap.

Historically, the development of the tidal mill parallels that of the river mill, but the obvious geographical limitations and the inherent difficulties in use prevented it from becoming a comparably important source of energy. One of the earliest British wheels was in operation in Wapping in 1233; it was owned by the Priory of the Holy Trinity at Aldgate. Many of these wheels operated successfully for centuries: one of the last to cease working was that at Pomphlett in Devon, which went out of action only in 1956.

The typical tidal mill was situated some distance inland on a tidal creek. The incoming tide flowed into a millpond, the gates of which were closed at high water. After the tide had ebbed, the gates were opened and the outflowing water was used to drive the mill.

The disadvantages are obvious. No two consecutive milling days could be the same because of the timing of the tide. And because of the variation in flow between spring and neap tides, a different amount of water was stored each day, thus altering the output of the mill. Also, since the output of any mill depends on the head of water driving it, the power output of a tidal mill varied throughout each period of operation, being greatest when the pond was full and declining as it emptied. Much ingenuity went into coping with these problems. Many mills used two or more ponds to store water for use during low tides or to boost output as the water level dropped in the main pond. A variable mode of operation was also developed, with sluices being used to make the wheel operate as overshot, breast, or undershot, depending on the water level.

Much the same limitations apply to modern tidal power installations. The first of these, and still the biggest, is at Rance in France. The average tidal range there is 8·4 metres. The area of the storage basin is 22 square kilometres and the maximum output capacity of the station is 240 megawatts. This is small by modern power-station standards. And whereas other kinds of power station can operate continuously, the tidal station is only theoretically available about a quarter of the time, and its peaks of power output vary, not in accordance with loads on the electricity supply system, but in accordance with the times of the tides. These disadvantages can only be overcome by the use of elaborate secondary, or even tertiary, storage basin systems. Although the 'fuel' is free, the low utilization of heavy capital investment makes a tidal power station compare unfavourably with conventional alternatives.

Nevertheless rising fuel prices and diminishing supplies are once again having the effect of making an unpromising method look more attractive. The Severn Estuary in Britain and Passamaquoddy Bay on the US–Canadian border have been the subject of intermittent interest for many decades and are now being examined again. Russia has a small experimental unit in operation at Kislaya Inlet. Various calculations have been done on the energy which might be obtained from tidal stations of this kind. According to Hubbert, development of all the world's favourable sites would yield an approximate generating capacity of 13 000 megawatts, about 15 per cent of the UK's total generating capacity, or 1 per cent of the total potential hydro-electric power.[17] It has been estimated that only about 25 sites throughout the world meet the necessary requirements of tidal range and coastal topography. The use of tidal energy is clearly limited.

The efforts of inventors to make use of the motion of the waves have been even less rewarding. One idea which, charming though it is, will not solve the problems, was put forward by the United Kingdom's Central Policy Review Staff. They suggest that a system of 'wave-generators' [sic] over a length of 900 miles of coastline would be able to produce about 30 000 megawatts, or sufficient to meet half the country's electricity requirements. The proposal deserves quoting if only for its optimism:

> The technology required for wave-generation is not expected to be particularly sophisticated. It could, for example, take the form of a system of floating tanks located about a mile offshore, tethered to each other and to concrete blocks on the seabed. The 'up and down' motion caused by the waves would be used to drive a high-pressure water pump with the resultant high-pressure water being used to drive a turbine

generator located on the shore. A 60-megawatt installation would extend to a length of about 1·7 miles and the tanks would have a height above water of about 4 feet . . . Wave-generators could not be installed immediately and some development work would have to be undertaken, but lead times would be comparatively short. Wave-generators would certainly have objections on aesthetic grounds and could cause some hindrance to coastal shipping. There would, in addition, be the danger that floats might break loose.[32]

This, with its nonchalant dismissal of problems and attention to irrelevant detail, is more the stuff of those dotty schemes with which patent offices are inundated than a serious proposal for supplying the country with electricity. Any harbour engineer or offshore oilrig constructor knows the force of winter storms at sea. Floating tanks would not bob up and down operating high-pressure pumps if they were merely tethered to each other and to concrete blocks on the bed of the sea. They would need to be as securely anchored as lighthouses or oilrigs. The effect of waves up to a hundred feet high streaming along a string of floats tethered to each other scarcely bears imagining. The design, construction and servicing of float guidance systems and seabed pumps similarly poses huge engineering problems. The timing of pressure pulses and their control in relation to each other is anything but an unsophisticated matter. The hazards to shipping caused by the installations themselves, or the floats which may break away, cannot be dismissed merely by recognizing that they exist. If there is one comment on this proposal that can be made with complete certainty, it is that the development time of 'wave-generators' will be anything but 'comparatively short'.

With the aim of getting 'away from the idea of an object bobbing up and down', Dr Stephen Salter of the University of Edinburgh proposed a different solution.[33] It would use the to-and-fro movement of the waves to drive a series of rocking vanes. He envisages a battery of vanes, built into a large floating structure, which would generate electricity to be used immediately to produce hydrogen by the electrolysis of water. He suggests:

The installations could be self-propelled. They could move line-ahead, a low drag condition, out into the Atlantic, turn abreast to the waves and be driven slowly back by wind and wave thrust, storing hydrogen on the way. Most of the hydrogen would be discharged at a shore terminal, leaving enough to get to sea again.

From these rather unpromising beginnings a more serious programme of research, almost entirely within the UK, has begun to emerge. It was launched in 1976 with a budget of £1 million and the allocation of funds has been increased since then. Some collaborative work with Japan is also being carried out. The various devices being investigated include a development of Dr Salter's rocking vanes, or 'ducks'; rafts hinged at the middle with pistons and pumps on either side of the hinge; columns of water oscillating up and down inside massive tubes; and pressure-sensitive devices sitting on the seabed. All these would require large structures; they would have to work in a very hostile environment; and they are at a very early stage of development. Preliminary estimates of the cost of electricity are that it might be up to twenty-five times as expensive as that from conventional power stations at present fuel prices. Such estimates are, however, premature. While ocean waves do contain substantial amounts of energy, the technology for extracting it simply does not exist. A proper evaluation of the potential of wave power requires another decade's intensive research and development.

An interesting idea for harnessing the energy of waves breaking on a coral reef has been worked out in some detail by Noel Bott, former general manager of the Central Electricity Board of Mauritius. An impounding lagoon is built behind the coral reef. The high waves coming in from the Indian Ocean break on a slipway and run up into the lagoon which fills to a level of 1½–2 metres above sea level. The water in the lagoon is then used to drive a series of pumps which lift water to the top of a nearby hill, whence it can be run down to generate electricity. Although the output is small this is an attractive proposal which uses nothing but well-tried and simple technology: it could find applications in suitable locations.

A completely different approach to taking energy from the oceans is to exploit the thermal gradient which exists in the tropical deep oceans. In some areas the temperature difference between the surface and depths of about 1 000 metres is as much as 25°C. In theory this difference could be used to drive a heat engine. The possibility was suggested by the French physicist Jacques d'Arsonval as long ago as 1881, and a 22-kilowatt trial plant was constructed in Cuba in 1919.

In the US proposals for gigantic machines are being studied. The companies involved, such as the Lockheed Missiles and Space Company, have experience in the outer reaches of military and space technology. The proposals rely on using the temperature difference to evaporate and condense ammonia as part of a cycle in which huge turbines are driven to generate electricity. The proposed process has acquired the name 'ocean thermal electric conversion' (OTEC). Again the unsolved structural, technical and maintenance problems rule out any hope of a significant energy contribution in this century.

Geothermal Power

The precise origin of the earth's internal heat is still debated. It is usually estimated that about 80 per cent is generated by the radioactive decay of uranium, thorium and their derivatives, with the remaining 20 per cent being residual heat from the original formation of the earth. The rate of heat-flow is tiny: about 0·0015 kilowatt-hours per square metre a day. The problem with geothermal power is the familiar one of concentrating a highly diffused energy flow into usable amounts. The fact that the temperature rises at an average rate of about 10°C for every kilometre of depth into the earth is apt to delude people into thinking that all that needs to be done is to drill a hole deep enough to reach the boiling temperature of water, pour water down it, and use the resulting steam to drive turbines. It is not as simple as this. While the temperature rises with depth, the flow of heat remains almost the same: there is no more energy flowing through a square centimetre 10 kilometres down than there is at the surface. Water poured down a 10-kilometre-deep hole would therefore rapidly cool the surrounding rock by withdrawing from it its stored heat. After a first flush of energy recovery the maximum rate of heat extraction would drop down to the slow rate of natural heat-flow through the rock. If geothermal energy is to be usable this natural heat-flow must be increased and concentrated.

The solid rocks of the earth's crust are usually about 30 to 40 kilometres thick in the land areas. At the base of the crust is the Mohorovicic discontinuity; below this is the mantle in which the rock is in a molten or semi-molten state. In some areas the mantle pushes upwards into the crust, increasing the local rate of heat-flow. Sometimes the mantle breaks through to the surface in areas of volcanic activity; at other times it merely creates a local 'hot spot' detectable only with sensitive thermometers or aerial infra-red surveying.

If this increased rate of heat-flow is to create a usable energy resource it must occur in an area in which the geological conditions are favourable to the formation of a geothermal reservoir. This is similar to the rock trap within which oil or natural gas accumulates: a permeable reservoir rock capped by an impermeable stratum. Ground water which has been heated by contact with hot rocks percolates into the reservoir where it becomes trapped beneath the cap-rock. If there are fissures in the cap-rock some of the hot water, steam, or both, may escape to the surface, forming hot springs or geysers. In such cases the underground system is replenished by further infiltrations of ground water. In nature the rate of energy loss is usually closely matched by the rate of energy gain and the system can remain active for a long time, though activity gradually diminishes. Geyser basins have a life of up to ten thousand years; lower-temperature hot springs last considerably longer.

The use of natural hot springs goes back a long way. The Romans made systematic use of them for medicinal and recreational purposes, as well as domestic heating. Appropriately enough, the Italians have been the pioneers of the modern use of geothermal energy for electricity generation. The Lardarello plant in northern Italy started producing electricity in 1904 and the total generating capacity there is now nearly 400 megawatts. The United States and New Zealand also have large installations generating electricity from geothermal sources. Table 27 gives a list of the countries which have electricity-generating stations using geothermal power: in all these countries further developments are being carried out.

Table 27. World installed geothermal electricity-generating capacity

Country	Generating capacity megawatts (MW)
El Salvador	60
Iceland	3
Italy	398
Japan	77
Mexico	75
New Zealand	192
USA	599
TOTAL	1 414

Source: *UN Statistical Year Book*, 1978.

Natural geysers or hot springs can rarely be used without drilling a borehole into the reservoir to control the steam or hot water flow. With present drilling technology this precludes the use of any reservoir more than 10 kilometres below the surface. When used for electricity generation the mineral properties of hot springs, which give them their curative reputations, are a real nuisance. The waters are rich in minerals because hot water under pressure is a powerful solvent; when the pressure is released, however, the water flashes into steam and the minerals are deposited on the sides of the borehole and reduce its diameter. Boreholes which supply power stations need to be re-drilled every seven to ten years. Other difficulties are caused when large quantities of these mineral-rich waters are run to waste from the power station: they can be extremely damaging pollutants in natural water courses.

The life and power-generating capacity of a geothermal reservoir obviously depend on the rate at which energy is extracted from it. Both are limited. The 1974 World Energy Conference survey says that '... the maximum capacity for the largest hydro-geothermal resources, when totally exploited over 50 years, is approximately 1 000 megawatts, and most large systems have capacities in the 100- to 500-megawatt range'.[27]

It is difficult to make any reasonable estimate of total world geothermal resources since so little exploration has yet been done. There appear to be belts of geothermal reservoirs along the western side of the Americas from Alaska to Chile; in Kenya, Uganda, Zaire, Tanzania and Ethiopia; in the Philippines, Indonesia, Burma and India; and in the countries around the Mediterranean (particularly Turkey). Hubbert[17] quotes calculations which indicate a possible world potential of 60 000 megawatts, about the same as tidal power, or the present total electricity generating capacity of the UK.

There is also interest in the possibility of obtaining heat from deep-lying 'hot-dry' accumulations of heat. These occur in rock strata where there is insufficient water present to carry the heat to the surface through boreholes. If the rock could be fractured sufficiently, however, water could be pumped down to capture the heat and bring it up. While experiments are continuing, extraction of geothermal energy by this method is not expected to be early or widespread.

The transformation of geothermal energy into electricity is at present the most usual approach. Because steam temperatures are generally low by modern power-station standards the generation efficiencies tend to be extremly low, often down round 10 per cent. This means 90 per cent of the heat energy is run to waste. Where the heat can be used directly for domestic heating as in Iceland and in some recently constructed housing estates in Paris, the generating loss is avoided and a much more efficient use is made of the energy. It is also possible to obtain useful amounts of energy from low-temperature geothermal sources. Another possibility being investigated in the UK is the use of such low-temperature sources for pre-warming the feed-water for power-station boilers.

Much of the technology of the oil industry can be employed in the development of geothermal resources. It is not surprising that the oil companies, as part of their policy of becoming 'energy companies', are among the major proponents of geothermal development. In California a minor rush of exploratory drilling, reminiscent of the early days of oil exploration, is already under way. In other areas of the world, countries hit by the increase in oil prices are turning to an examination of their geothermal resources. Kenya is an example among developing countries. It looks, however, as though geothermal energy, while it may be important in favoured areas, will remain a minor energy source in a global context.

Wind Power

Windmills are perhaps the most pleasing of all devices for harnessing energy – and the only ones which embellish an inhabited landscape. They are reputed to have been used in China, in a vertical-axis form, more than 2 000 years ago. In the West they appear in Arab writings of the ninth century. The first recorded northern European mills were in France in 1180 and in England in 1191.

Early windmill technology was often derived from that of the waterwheel, but in time a distinctive response to the very different problem of extracting energy from the wind was evolved. Air has a density nearly eight hundred times less than that of water. Therefore, to capture the same amount of energy from the same speed of flow, the windmill needs eight hundred times the blade area of the watermill. The windmill's requirements of size, lightness, robustness, and adaptability to varying speeds and directions of wind made heavy demands on the ingenuity and skill of its designers.

Their solutions frequently combined considerable technical and aesthetic flair.

The windmill developed steadily over the centuries. Means of braking and feathering the sails and of remote control, improvements in gearing and power transmission combined to give it a role in the medieval economy comparable with that of the watermill. A particularly ingenious innovation was the fantail. Before its invention the difficult and dangerous job of keeping the sails facing into the wind had to be done by hand. The fantail was a small vaned wheel, another windmill in fact, mounted at right angles to the main sails, behind the cap of the windmill. When the wind veered out of alignment to the main sails, it began to turn the fantail. This operated a gearing arrangement which slewed the main sails back into the wind. This early example of a negative feedback device made a great difference to the safety and efficiency of windmills. It was patented in England in 1754 by Edmund Lee.

Large windmills could develop up to 30 kilowatts, which was enough for the milling needs of a small community. Holland, the country most famous for its windmills, had 8 000 of them in the middle of the eighteenth century. But Germany had 18 000 in operation in the nineteenth century; Britain had 10 000, mainly in the windy south-eastern counties; and Portugal had 1 000 operational as late as 1965. Small windmills, or 'windchargers', were a familiar sight in rural areas a couple of decades ago. These generated a small amount of electricity which was stored in batteries and used to operate domestic lighting circuits and radios. These machines are also used for water pumping, especially for irrigation. Many are in use today in California to provide water for the orange groves.

The sail windmill in its developed form was a beautiful solution to a very difficult set of problems. It is an immense tribute to its designers that it could in many cases compete with the watermill. Both, however, were the end of a line of technical development. The waterwheel was completely superseded by the much more efficient water turbine. The analogous development in wind power is the propeller mill. Most modern windmills are of this type, which rotates more quickly and is a generally more efficient energy collector than the sail mill. Unfortunately, no progress in the general development of wind power comparable with that which has occurred in the case of water-power technology has taken place. Since the winds cannot be manipulated, dammed or pumped, like water, windpower has been relegated to a very minor role in the important business of generating electricity.

Most of the early work on generating electricity from wind was carried out in Denmark, starting at the end of the last century and continuing into the 1940s. Machines capable of generating up to 70 kilowatts were built. During the 1930s the Lucas 'Freelite' machine was on the commercial market; this could power four or five light bulbs of about 40 watts each. Similar machines were on sale in the US. A 30-metre-diameter machine delivering 100 kilowatts was built in Russia in the 1930s.

The largest machine of all was built during the 1940s in the US at Grandpa's Knob in Vermont. This had a two-blade propeller 52·5 metres in diameter and delivered a maximum of 1 250 kilowatts. Wartime difficulties with spare parts and finally an accident in which one of its blades broke off during a storm ended the life of this machine. Just after the war there was a surge of interest in Britain; a 100-kilowatt machine was built at St Albans and another in the Orkneys. Enthusiasm for these experiments was not sustained as electricity generation using coal or oil was then so cheap.

The perennial problem of keeping a windmill facing into the wind has led to the design of various vertical axis windmills. These include the Savonius rotor which, in its simplest form, can be made by cutting an oil-drum vertically in half and welding the two parts together to form an S-shape; and the Darrieus rotor, which is simply a hoop with an air-foil section mounted on a vertical axis. These machines are not self-starting, a major disadvantage which must be set against their other advantages.

Contemporary research effort is mainly in the US where contracts have been awarded to companies such as Boeing and Grumann, better known for their involvement in aerospace and military technology. They are investigating large and sophisticated machines which might be fitted into electricity grids. In the UK it has been estimated that giant machines constructed in shallow offshore waters could contribute up to 10 per cent of the country's electricity needs by the year 2000.

Because of the low energy-density of wind, machines have to be large, light and strong. For the designer these are conflicting requirements which can only be met by compromise. It also means that the maximum output of practicable machines is unlikely to exceed 1 000 kilowatts – 1 megawatt. Most modern power stations are 1 000 megawatts; replacing even one of these by a thousand huge wind-machines is almost unthinkable. If this were not limitation enough, the variable output and unreliability of windpower further diminish its prospects for use in large-scale power generation. Small machines, on the other hand, are potentially useful, especially in rural areas. In the developing world, where electricity grids do not exist, or where electricity is prohibitively expensive, windmills of modern design may have some useful role to play in water pumping and electricity generation.

Wood

Throughout history wood has been the fuel on which the majority of people have relied. It still is. Having almost entirely given up the use of wood as fuel, industrial societies tend to forget that in the developing world wood is still the principal source of energy for billions of people. Accurate data are, of course, hard to get; not many of the people concerned fill in statistical returns. But in countries such as Thailand, Gambia, Tanzania and many others, wood supplies 90 per cent of the fuel needs of the whole population. In other developing countries, such as Kenya, where oil supplies a somewhat higher proportion of total energy needs, wood remains almost the only fuel in rural areas.

It has been estimated that annual world consumption of wood as fuel is about 500 million tonnes of coal equivalent, about 6 per cent of the world's total energy consumption. But almost everywhere depletion of forests proceeds faster than replenishment. There is a merciless and frightening arithmetic of population growth and disappearing trees which is propelling hundreds of millions of people towards disaster. The appalling suffering in the Sahel region of Africa is an example. There, loss of forests and the encroaching desert have left people with neither fuel nor food.

The obvious answer would seem to be intensive reafforestation programmes and some countries are attempting them. Forest renewal is not easy and there is usually competition with other land uses. There is also competition for the wood produced, from the building and paper industries in particular. Thus a new problem arises: wood, instead of being gathered free, has a price. But incomes are so low that money for fuel is simply not there.

Alternative fuels cannot be introduced for the same reason. In other words, incomes must first be raised so that people can afford to buy, say, kerosene and a stove to burn it in. While cheap oil appeared to be available in unlimited quantities, the depletion of fuelwood could be viewed almost with equanimity: in time the transition from wood to oil would be made as countries developed and prosperity spread. Now it is clear that this will not happen – at any rate, not in time. There is no simple approach to these interdependent problems. The richer nations must appreciate the gravity of the fuelwood shortage, and they must be prepared to offer all the help they can to avert a real catastrophe. For wood, over a rapidly increasing area of the world's surface, is likely to be the first major energy source to 'run out'.

Ironically, some of the richest countries in the world are amongst the most well-endowed with forest resources. For them the possibility of relying on wood for a great deal more of their energy needs than they do at present is beginning to look increasingly attractive. In the big timber-producing countries of the morthern hemisphere, Finland, Canada and Sweden for example, wood and wood wastes from pulp and sawmills already provide sizeable amounts of energy. Ambitious proposals have been made for much greater reliance on wood as a renewable fuel in such countries. The study *Solar Sweden*,[34] for example, gives detailed calculations demonstrating the theoretical possibility of meeting all Sweden's future energy needs from renewable sources. A major contribution would be made by 'energy plantations'. These used to be called forests. In their new guise they are simply areas of woodland carefully cultivated for maximum yield. The wood would be burned or processed into liquid fuels. Under such a system the total yield of energy on a sustainable basis, and bearing in mind the need to replace soil nutrients, might be about 90 000 kilowatt-hours per year per hectare.

While such schemes might appear to offer good prospects of sustaining industrial life in an era when oil and coal are extremely scarce, their hypothetical nature and the very particular circumstances under which they might be realized must be remembered. Sweden and North America have about 2 hectares of forest per head; Africa has about 0·5, and Asia only 0·2. Moreoever, in the developing countries much of the remaining forest is far from the areas of greatest need. In the more densely populated industrial countries there is simply not enough land; meeting the UK's present energy consumption from such energy plantations would require that the entire area of the country be given over to the cultivation of trees.

It has been estimated[35] that the total energy contained in the world's present forest stock is about 270 billion tonnes of coal equivalent; this is about two-thirds the figure for ultimate reserves of oil. The annual addition by growth to this is about 7·7 billion tonnes of coal equivalent. In a stable forest system this growth is matched by an equal amount of decay; much of it, however, could be harvested on a sustainable basis, provided the integrity of the forest was maintained and the necessary nutrients were returned to the soil. Only about 15 per cent of the increment is, in fact, harvested at present, roughly half for fuel and the rest for industrial purposes. Genetic improvements in trees and plants could certainly increase the amount of wood produced. But the irony and the tragedy remain. Where the need for fuel is greatest the forests are vanishing; for the hundreds of millions of people concerned it is an entirely academic matter that wood is abundant elsewhere.

12

Energy Conversion and Storage

Electricity

Electricity came into commercial use towards the end of the nineteenth century. Coal had been the foundation of the Industrial Revolution, making energy available in quantities for which there was no historical precedent. But it was an awkward fuel, dirty, unwieldy, difficult to transport and store, and inefficient in use. Electricity overcame most of these difficulties.

It was clean. Transport was hardly a problem: it could be transferred almost instantaneously. Because it required neither handling nor storage it was economical in labour and space. Once the basic installation had been provided, electricity was an ideal form of energy. Many new uses were devised for it. Electric light rapidly superseded all other forms of illumination. Electric motors could be made tiny enough for the most delicate tasks, or robust enough for heavy industrial use. Safe, clean urban transport systems could be powered by electricity; domestic drudgery could be reduced by using cookers, refrigerators, vacuum cleaners and washing-machines for clothes and dishes. Electricity had a major part in shaping the industrial economies of the twentieth century. Table 28 shows how its use has grown since 1929, doubling every ten years or so.

The first electric power station commenced operation in London on 12 January 1882; the next began in New York on 4 September of the same year. Problems with payment of electricity bills must have been immediate. People who do not pay their electricity bills in Britain today are disconnected under the provisions of the Electric

Table 28. Growth of world electricity production – 1929–77

Year	Production gigawatt-hours (GWh)
1929	(287 300)*
1935	(333 000)
1940	(482 000)
1945	(572 000)
1950	(858 000)
	956 800
1955	1 535 500
1960	2 301 000
1965	3 377 400
1970	4 923 400
1973	6 087 100
1977	7 209 682

Source: *UN Statistical Year Books.*
*Figures in parentheses exclude the USSR and China.

Lighting Act of 1882. These early supply systems were small and primitive, using direct current transmitted at a low voltage. Service areas were limited by the length over which current could be transmitted economically, which was only a few kilometres. Transmission losses made inter-city services quite prohibitively expensive. But technical progress was rapid and by the turn of the century the use of direct current had given way to alternating current and, in California, electricity was being transmitted over a distance of 110 kilometres at 40 000 volts.

High voltages brought problems with insulators, pylons, switchgear and ancillary equipment. But they reduced the costs of cables and cut down transmission losses.* They extended the range over which electricity could be distributed without becoming economically uncompetitive. By 1920 transmission voltages had reached

*For a given power the current is inversely proportional to the voltage. Raising the voltage therefore reduces the current while allowing the same power to be transmitted. Since transmission losses are proportional to the square of the current, increasing the voltage by a factor of two will reduce the losses by a factor of four.

132 000 in some places. After the Second World War there were further major developments in electrical technology. In 1954, Sweden, which has been very prominent in the development of electric-power generation and transmission, brought a 380 000-volt line into service. This connected the Harspranget power station inside the Arctic Circle with southern Sweden, a distance of 970 kilometres. In 1965 Canada introduced a 765 000-volt transmission line, and work is continuing on even higher voltage systems in the USSR, US and other countries.

The increased efficiency of transmission has allowed national or even international transmission grids to be created. These link the generating stations in an area into a common system. A central computer control monitors the changing load conditions in different parts of the grid and switches stations in and out as required. Areas of high load can be supplied from those where there is a surplus of generating capacity. The system can be arranged so that the modern, efficient stations are used as much as possible with low-efficiency stations only being brought in to supply short-term peaks. The larger the total capacity of the grid the greater its ability to deal with breakdowns. In a grid such as that in the UK, even if a large station breaks down suddenly its load is taken up within fractions of a second by the remaining stations, without consumers noticing anything. A disadvantage is that the effects of a major breakdown or a loss of central control will be transmitted with equal ease and speed over the whole of the grid.

World electricity production in 1970 and 1977 is shown in Table 29. Of the 1977 total 36 per cent was generated in the US, 27 per cent in Europe and 16 per cent in Russia. The rest of the world shared the remaining 21 per cent – a familiar pattern which has barely changed since 1970. A fifth of the total was generated by hydro and 7 per cent by nuclear power. The remainder was generated using fossil fuels and required an input of about 2 billion tonnes of coal equivalent, which was about 20 per cent of all the fossil fuels consumed in the world that year.

Much work has been done to increase the efficiency of both

Table 29. World electricity production – by regions, 1970 and 1977

Region	Production gigawatt-hours (G Wh) 1970	% of total	1977	% of total
Africa	87 400	1·8	147 520	2·0
North America	1 898 000	38·6	2 619 053	36·3
South America	107 600	2·2	197 828	2·7
Asia	612 200	12·4	1 013 836	14·1
Europe	1 407 400	28·6	1 973 177	27·4
Oceania	69 900	1·4	108 194	1·5
USSR	740 900	15·0	1 150 074	16·0
TOTAL	4 923 400		7 209 682	

Source: *UN Statistical Year Book*, 1978.

Note: Of the 1977 total 508 951 G Wh (7 per cent) were generated by nuclear power. About 1 510 000 G Wh (21 per cent) of the total from hydro power.

generation and transmission. Since the beginning of the century the efficiency of generation has been increased from round about 10 per cent to the present maximum of around 40 per cent in a large power station. This has been done mainly by increasing the operating temperature of the steam turbine. Present temperatures, of around 600°C, are close to the performance limits of reasonably available materials. Further gains will thus be hard to win.

Major improvements in efficiency have been obtained in the past by increasing the size of generating units. These are now commonly 600 megawatts or more. One of the disadvantages of units this size is the big change in the generating capacity of the system which occurs when they are brought in and out of use. The system needs to be large enough to damp this effect to an acceptable level. Larger units would exacerbate this problem. Possible future progress in this direction is therefore limited.

There is, however, a possibility of developing a new form of generation in which the turbine is eliminated completely. If an ionized high-temperature gas is passed at high speed through a strong magnetic field an electric current is generated and can be extracted by placing electrodes in the stream of gas. This is MHD, magneto-hydrodynamic generation. Although the principle was established by Faraday, many technical obstacles still remain to be overcome before large MHD generators are practical. The motive for developing them is strong. They could be employed as the first stage in the generation process at a power station. They would take very hot gases directly from the furnace or nuclear reactor at temperatures of up to 1 500°C, which is much too hot for use in conventional generating plant, and use these to produce some electricity. In the second stage, the gases, cooled to normal operating temperatures, would be used to raise steam and drive turbines in the usual way. This interception of the combustion gases and extraction of energy from them is called a 'topping cycle'. Although an MHD topping cycle might double the efficiency of electricity generation and is thus attractive there is little possibility of large units being commercially available before the turn of the century.

Another method of electricity generation with a high theoretical efficiency is the use of a fuel cell. Again the principle is old – the first fuel cell was made by Grove in 1839 – but development has been slow. Grove used a cell with two platinum electrodes immersed in a solution of sulphuric acid. By feeding oxygen to one electrode and hydrogen to the other he obtained an electric current and produced water in the cell. Passing an electric current through the cell reverses the process and electrolyses the water back into hydrogen and oxygen. Theoretically, efficiencies can approach 100 per cent but in practice there are difficulties. Despite a considerable amount of research fuel cells large enough for industrial generation of electricity have not yet been produced. Research is being carried out in the United States into a variety of types of fuel cells using different fuels. One of the most heavily funded areas of this research is into the use of oil or natural gas as the feed for the cell. The greatly increased efficiency of generation could justify the use of these fuels even at a time when they are being excluded from normal power-station use. Progress, however, is slow and fuel cells offer no hope of a significant contribution over the remainder of this century.

Super-conductors can cut transmission losses dramatically. When the temperature of a conductor is near absolute zero, transmission losses almost completely disappear. The resistance of copper is reduced to a five-hundredth of its normal value at a temperature of 20 K. Practical development of super-conductors could therefore lead to transmission systems of high capacity with very low losses, such as one suggested by Garwin and Matisoo of IBM which would be 600 miles long and carry 100 000 megawatts. Such technology is, however, 'still far in the future'.[36]

Electricity Storage

A major difficulty with electricity is that it must be used immediately it is produced. It cannot be stored directly. This is particularly irksome when there is little or no control over the energy source used for generation. Wind, solar and tidal power must be used when they are available or not at all. Nuclear power stations have great difficulties in following the rise and fall of electricity consumption throughout the day and night. Most of these difficulties would be eliminated if a cheap and effective means of transforming electricity into a storable energy form, from which it could be recovered at will, could be developed. This has been the objective of a great deal of research.

Transformation into chemical energy is the most common approach. Despite immense efforts, the lead–acid battery, as used in motor cars, remains the most practical device yet produced. These batteries use lead electrodes immersed in a solution of sulphuric acid. They need careful use and attention. They must be kept topped up with distilled water and not allowed to remain too long in a discharged condition. Rapid charging and discharging limits their life to a few hundred cycles. They are cumbersome and heavy; a battery to store a single kilowatt-hour weighs about 25 kilograms.

Other battery types which have been investigated include nickel–iron, nickel–cadmium, silver–zinc, zinc–air, sodium–sulphur and nickel–zinc, as well as more exotic high-temperature systems. None is decisively superior to the lead–acid battery. As Hottel and Howard say: 'The importance of good cheap storage batteries is so clear that there is no problems in stimulating research in this area.'[31] But useful results remain obstinately out of reach.

Another approach is to use surplus electricity to pump water into a high-level reservoir from which it can be withdrawn at periods of peak electrical load. Some large-scale pumped-storage schemes, as they are called, have been built and many more are being discussed in various countries. They can play a valuable part in balancing loads and stabilizing the output of a country's electricity system. But they are very expensive and as a reservoir needs special topography the number of suitable sites is limited.

To extend the possibilities of this principle, underground pumped hydro-electric storage systems are being considered. The lower reservoir would be constructed as an underground cavern in hard rock. As the drop between upper and lower reservoir can be much greater – several thousand feet – much smaller reservoirs are required for the same output.

Underground storage of compressed air could be even more advantageous. No surface-level reservoir is necessary and the underground caverns could in some cases be leached out of salt formations. A large compressed-air storage system on these lines came into operation recently near Bremen in West Germany; it can produce 290 megawatts for about two hours. Surplus generating capacity is used to compress air to about 1 000 pounds per square inch. At times of peak demand the air is expanded into a turbine. At the moment, the air has to be cooled after compression and reheated on expansion using an energy input, but a method of storing the heat of compression would obviate this.

Some investigations have been carried out into the use of flywheels to store energy. Huge flywheels weighing between 100 and 200 tonnes and spinning at 3 500 revolutions per minute could store between 10 000 and 20 000 kilowatt-hours.[37] Flywheels have the advantage of absorbing and releasing energy rapidly. But they are expensive. They are also extremely dangerous, as can be appreciated from the above figures. They would probably have to be housed underground, thus becoming prohibitively expensive.

Offpeak electricity could also be used to electrolyse water. When an electric current is passed through it, water breaks down into hydrogen and oxygen, a familiar experiment in junior science classes. The hydrogen thus produced can be stored and used as fuel for combustion or for electricity generation in a fuel cell.

Hydrogen has a great deal to recommend it as a fuel. It can be piped, stored in gasholders, or liquefied and stored under pressure. It can be burned in cookers, furnaces, motor cars and aeroplanes. It is clean – the only by-product of its combustion is water. Its only obvious disadvantage is its comparatively low calorific value which is about a third that of natural gas. But it can be combined with carbon, obtained from the almost limitless supplies of limestone, to form methanol, a liquid fuel with most of the desirable attributes of petrol.

So great are the attractions of hydrogen as a fuel that it has led to the invention of the 'hydrogen economy'. This is an imaginary future in which hydrogen is substituted for coal, oil and gas; the developed world thus continues its present patterns of energy consumption indefinitely into the future and long after all the reserves of oil and gas have been consumed.

Hydrogen, however, is not an energy source. It is a manufactured fuel, like electricity; and like electricity its manufacture consumes more energy than can be obtained from the resulting hydrogen. In electrolysis about half the energy of the electricity used can be recovered from the hydrogen produced. Using coal to produce electricity which in turn produces hydrogen therefore results in a conversion efficiency of 15 to 20 per cent even under the best conditions. It can only be justified under exceptional circumstances, for instance, to produce hydrogen for use in chemical processes. In fact, hydrogen is more usually produced for this purpose by breaking down natural gas.

Obviously the hydrogen economy would not use fossil fuels to produce its hydrogen; using them directly is far more efficient.

Neither would it use electricity which could be used directly. The hydrogen economy would rely on using a large quantity of electricity which would otherwise go to waste. It depends therefore on two assumptions about the future of nuclear power. The first is that the efforts to develop nuclear power will be extremely successful: not only will nuclear power be able to meet all the conventional requirements for electricity but it will be capable of generating a very large surplus. The second assumption is that nuclear engineers will be extremely unsuccessful in the attempt to build stations which can follow the changes in electrical load easily: if they succeed in this it will be possible to use electricity directly for most uses without the expense and energy waste of turning it into hydrogen. The hydrogen economy rests on the unlikely hypothesis of the simultaneous fulfilment of these two assumptions; until it is much more securely based it can safely be ignored.

This is not to say that using electricity to produce hydrogen has no future whatsoever. If a cheap and simple electrolysis 'kit' could be produced, the intermittent and unpredictable output of a wind-powered generator could be stored and the surplus energy of a solar-powered generator could be slowly accumulated during the summer and used in winter. On a larger scale, electrolysis could be used to serve the same purpose as today's pumped-storage systems and store some of the energy which would otherwise be wasted at nuclear, hydro or tidal power stations and use it to help meet peak loads. But the evolution of these uses into a full-scale hydrogen economy is a matter for the distant future, if at all.

Another long-term possibility is to use the heat from nuclear reactors to produce hydrogen directly from water, without an intermediate conversion to electricity. This could have a great deal to recommend it in terms of efficiency, at least in theory – it should be possible to devise a process with an efficiency of considerably more than the 15 to 20 per cent obtained by first using the heat to produce electricity and then using the electricity to produce hydrogen. But there are still large difficulties to be overcome before this can happen, as can be seen from the following quotation:

The scheme to which most attention has been given is based on reactions between $CaBr_2$, $HgBr_2$ and water. While many of the basic chemical parameters have been studied, a very large chemical engineering development programme would be needed . . . criticisms which have been levelled at this scheme relate to the environmental implications of handling mercury on such a scale, the mercury investment required and the corrosion problems inherent in such corrosive materials.[38]

Gas from Coal

Another way in which coal can be made more convenient to use is to turn it into gas. This has a long history. In Europe, coal-gas was used for illumination in the late eighteenth century. The London and Westminster Gas Light and Coke Company was granted a charter in 1812 and the first US company was chartered in 1816 in Baltimore.

The basic process was very simple. Coal was heated in airtight retorts to drive off its volatile constituents. About 70 per cent of the original coal was left behind as a coke of almost pure carbon which was used for smelting and as a furnace fuel. The gas produced was a mixture which depended for its exact characteristics on the coal from which it was formed and the type of process used.

In Table 30 a typical range of constituents in gas made from the Durham coalfield in the UK is shown. Over half the gas is hydrogen, with a fifth to a third being methane. The carbon monoxide is also extremely important: it is the lethal constituent. When inhaled it reacts with the haemoglobin in the blood, preventing it from carrying oxygen from the lungs. The introduction of non-toxic natural gas was welcomed, amongst other reasons, for discouraging 'easy' suicides.

Another coal-gas with a long history of manufacture is 'producer' gas. This is made by passing steam and air over an incandescent bed of coal. The steam and the oxygen from the air combine with carbon from the coal to give the combination of gases also shown in Table 30. Here, the main combustible constituent is carbon monoxide, the carbon dioxide and the nitrogen being inert gases. As the calorific value of producer gas is only about a third that of town gas, and as it has a high content of the toxic carbon monoxide, its use is limited to specialized industrial processes.

Table 30. Constituents and calorific values of town and producer gas

CONSTITUENTS percentage	Town gas*	Producer gas†
Carbon dioxide	1·6–5·2	5–10
Oxygen	0·3–0·5	—
Unsaturated hydrocarbons (ethylene, propylene, butylene)	2·0–4·0	—
Carbon monoxide	6·2–16·4	25–30
Hydrogen	52·9–55·3	10–18
Saturated hydrocarbons (methane, ethane)	19·2–30·2	0·5–3·0
Nitrogen	5·1–7·4	50–60
CALORIFIC VALUE	5·17 kWh/m³ (500 Btu/ft³)	1·34–1·55 kWh/m³ (130–150 Btu/ft³)
YIELD OF COKE PER TONNE	0·655–0·755 tonnes	—

Source: E. C. Pope, ed. *Coal: Production, Distribution, Utilisation* (Institute of Fuel and Coal Industry Society), Industrial Newspapers, 1949.

*Constituents of town gas are those obtained from a Durham coal processed in a variety of ways.

†Constitutents of producer gas are those obtained from processing a range of fuels from coke to anthracite.

Most gas-works producing town gas or producer gas have now closed. The superior calorific value of natural gas, twice that of town gas; its non-toxic and non-pollutive characteristics; its cheapness, as an unwanted by-product of oil production; and its ready availability, have led to its complete dominance of the United States gas market since the Second World War. In Britain a similar development occurred after the discovery of the southern North Sea gasfields in the early 1960s and most of the old town-gas distribution network has been modified to take natural gas.

Gasification of coal is again receiving attention, particularly in

America where supplies of natural gas are declining. A variety of approaches is being explored. The simplest method is to produce a gas of low calorific value containing methane, hydrogen and carbon monoxide, by heating coal in the presence of steam and air or oxygen. The resultant gas, resembling producer gas, is cleaned of impurities and used either for industrial processes or power generation. This kind of gas, because of its low heating value, is uneconomical to transport over long distances; it is therefore used where it is produced. In effect, the process is a means of making coal more convenient to use.

To produce a gas of high calorific value, sometimes called substitute natural gas (SNG), further processing steps are required in which the hydrogen and carbon monoxide pass through a catalytic reaction in which methane is formed. The resultant gas can be used either as a chemical feedstock or for upgrading to a liquid hydrocarbon, or for distribution through a natural gas network.

The technical problems of making gas from coal have all been solved. Given a commitment to doing so, there is no reason, in principle, why large-scale gasification of coal could not be taking place within the next decade. There are, however, economic and engineering difficulties to be overcome. Although a lot of money is being spent on research and development, particularly in America and Germany, neither governments nor private industry have yet been prepared to move into the construction of large commercial plants. The exception is South Africa. There, a combination of political isolation and extremely cheap coal have encouraged the development of coal gasification to a level not found elsewhere. The technologies in use are, however, generally based on those used in Germany during the Second World War and unlikely to prove economically attractive to many other countries.

There is little doubt that commercial gasification of coal will come into operation over the next couple of decades. Different processes will be used, depending on the product gases required and on the kinds of coal available as feedstocks. But a price will be paid for the convenience of continuing to have gas available for industrial and domestic use. To feed the gasification plants there will have to be an increase in coal mining with all that implies in environmental damage and disruption. Replacement of natural gas by SNG is possible; no one should be under the illusion that it will be easy or cheap.

The underground gasification of coal, which would eliminate the problems of its mining and transport, was first suggested by Sir William Siemens in 1868 and at various times since then trials have been made but without much success. To make the gas, a series of holes is drilled into a seam of coal, preferably one sloping regularly upwards. A controlled fire is lit within the seam, and air, oxygen, steam or some combination of these, is passed through the holes and into the combustion area where a kind of producer gas is formed and drawn off. This can be used directly, upgraded into town gas, or used as a feedstock for a methane-producing process.

The temperature and extent of combustion cannot be controlled with precision, yet these are critical in the formation of the gas. Variations in the quality of the coal further complicate the control problems. The output gas varies widely in its characteristics and, if air is used as feed, has a calorific value only about a tenth that of natural gas. And, of course, most of the carbon in the coal is either consumed in the combustion or left in the form of coke below ground. Although there has been intermittent research in a number of countries and there is still some interest in Russia few expect underground gasification to be developed within the next few decades.

Oil from Coal

It is well known that Germany produced considerable quantities of oil from coal during the Second World War. The process was not in any way competitive with oil from conventional sources; it was developed only as an indigenous source of fuel to safeguard Germany against a blockade of external supplies. South Africa is the only country which has seriously developed and kept in operation the German process. A plant producing 4 000 barrels a day (200 000 tonnes a year) is in operation there at Sasolburg. It produces both gasoline and diesel fuel and uses about 4 400 tonnes of coal a day. This plant is at present being augmented by another with a capacity of 1·5 million tonnes of liquid fuels per year.

A number of other methods of liquefaction are now under development, mainly in the United States. The processes are complex and the plant required is expensive. A common requirement of all methods of liquefaction is that hydrogen be added; without this it is impossible to form the complex hydrocarbon molecules of liquid fuels. Unfortunately, the production of hydrogen is expensive and energy-intensive.

The products of the different liquefaction processes vary widely. Some produce liquid hydrocarbons as a direct replacement for today's petroleum products; others produce a dense solid which becomes liquid only when heated. This solid fuel is, however, clean and non-polluting and can be used in power stations.

Liquefaction has been dogged by economic and technical problems. The estimated costs of producing liquid fuels have obstinately remained higher than those derived from crude oil, no matter what the price increases in the oil market. A major economic difficulty is that at the oil prices at which oil derived from coal is likely to be competitive the demand for oil falls as users turn to coal or begin to economize. This will tend to shrink the market for liquid fuels and eliminate the need for liquefaction. Such fears have prevented any of the companies engaged in research from committing themselves to the construction of full-scale plants.

The development of liquefaction is likely to be quite slow over the next decade. Later, depending on the oil market and the priority given to security of supply, liquefaction may begin to take up a slowly increasing share of liquid fuel production. But, again, the difficulties should not be forgotten. Net yields are unlikely to exceed 2½ barrels per tonne of coal. To produce the present oil consumption of the US from coal would require the mining of about 2·5 billion tonnes of coal a year – about the present production of coal in the whole world.

There are also fears about new forms of pollution. Among the complex hydrocarbon molecules formed in liquefaction there are several which may be potent carcinogens. The problem of controlling the emission of such products could be difficult and very expensive to resolve.

Before heavy commitments are made to ways of converting coal to oil it is more sensible to examine ways in which coal might be substituted directly for oil. It is absurd to convert coal into oil and then burn the oil if coal could be used equally easily as a fuel. Many oil-fired power stations, for instance, could be converted to burn coal with little difficulty and no loss in efficiency. This would release oil for those uses for which it is uniquely suitable. Over the next few decades this approach would do more to ease the pressure on oil resources than any possible development of processes for turning coal into oil.

Fluidized Bed Combustion of Coal

One of the most promising of the newer coal technologies and which could greatly help extend its use is fluidized bed combustion. This had its technical beginnings in the 1920s but, as happened with so many of the initiatives of the early decades of the century, the Second World War and the arrival of cheap oil killed commercial interest. In recent years research has been resumed in the US, Britain, and Germany.

In a fluidized bed combustion unit air is passed upwards through a layer, or bed, of fine incombustible material, causing it to behave as a fluid. Crushed coal is fed into this fluidized bed and burned there. The combustion temperature is about 850°C; this is just about half the temperature in a normal boiler. Boiler tubes run directly through the bed and absorb the heat required to generate high-pressure steam. Fluidized bed units can be designed to run at atmospheric or at elevated pressures.

Fluidized bed combustion, contrary to some of the exaggerated publicity it has received, introduces no revolutionary principles nor does it radically change the energy outlook. Nevertheless, it does have some important advantages over conventional coal-burning boilers. The transfer of heat to the boiler tubes is more efficient than in a normal boiler where the tubes are in the walls and roof of the combustion chamber. This increased efficiency of heat transfer means that for any particular heat output the fluidized bed unit can be smaller than the conventional boiler. It is hoped that units will be produced of the same size as oil-fired units of the same output; this would increase the possibility of replacing existing oil-fired boilers with fluidized bed installations.

The quantity of inert matter contained in the coal burned in a fluidized bed can be much higher than would be acceptable in other boilers. This permits the use of lower and cheaper grades of coal; in fact, much of the colliery waste piled in the slag heaps beside coal-mines can be burned quite easily in fluidized beds. Another advantage is that when limestone or dolomite is added to the bed up to 90 per cent of the sulphur contained in the coal is absorbed; this greatly reduces the problem of pollution.

A further bonus is provided by the fact that the lower combustion temperatures reduce the formation and emission of oxides of nitrogen. These oxides, which are the main cause of atmospheric smog, are likely to be made subject to regulation in some countries. In conventional boilers, because of the high temperature at which they operate, atmospheric nitrogen is combined with oxygen, as happens in the engine of an automobile. In a fluidized bed unit, because of the lower operating temperature, no oxides of nitrogen are formed in this way.

A small fluidized bed unit has been developed in China and hundreds of these are in use. They use low-grade coals or rubbish, and appear to be mainly used for heating rather than electricity generation. Demonstration units, which are hoped to be the precursors of large commercial units, are in operation in the UK and the US. A large unit, jointly funded by the US, West Germany and the UK under the sponsorship of the International Energy Agency, is expected to come into operation in the early 1980s.

Although the principles of fluidized bed combustion are familiar and small-scale units for heat raising can be built relatively easily, advanced fluidized bed technology still faces many problems. There is considerable pessimism among some experts about the rate at which the technology can be brought into widespread commercial use. Power stations using fluidized bed burners are unlikely to come into widespread use much before the turn of the century; small-scale industrial and commercial applications may, however, become feasible somewhat earlier.

Methane and Alcohol from Biomass

There is nothing new about biomass except the name. The solar energy trapped by photosynthesis is usually used directly by man as food for himself or his animals. It is also used directly as fuel when wood is burned. Other less direct ways of using plant material include anaerobic digestion to produce methane, and fermentation to produce alcohol. 'Biomass conversion' is coming into use as a term to describe the production of gas or liquid fuels from plant material; it may, however, be used to describe the general use of plant material for fuel.

When organic material decays under anaerobic conditions, that is without oxygen, as for example in the mud at the bottom of a marsh, methane is produced. This is the marsh gas which produces the mysterious will o' the wisp if it ignites. In municipal sewage works where the organic material of sewage is digested in sludge digestion tanks, methane is also produced, and sometimes referred to as sludge gas. The technology of sludge digestion is well established and the gas is often used to produce power for the sewage works. Sometimes there is even a surplus which, as one authority on sewage plant design says, can be 'burnt in a burner at some distance from the tanks. This can be designed as an attractive ornamental feature.'[39]

In the late 1960s methane production was very much in vogue and widely publicized. It was even referred to as the 'fuel of the future'. It became the subject of much earnest but amateur experimentation which, luckily for the experimenters, was generally a failure. Air and methane form an explosive mixture. The only reason many of the leaky home-made digesters did not cause fatal accidents was that they rarely produced any methane. Effective operation demands fastidious care and control of the temperature and the mixture of materials used.

As a substitute for natural gas as it is presently used in the industrialized countries methane produced from sewage has no future. The net yield from sewage works is negligible and small-scale domestic digesters, even if fed with plants specially grown for the purpose in addition to household wastes, are little better. In cold climates, or during winter, keeping the digester at the required temperature will generally consume more energy than is produced.

The production of methane on farms, particularly if they use battery methods of animal rearing, is more promising. The quantities of animal and vegetable wastes are usually sufficient to justify a reasonably large digester, and their disposal by other ways can be difficult and expensive. A methane digester could provide a high proportion of the energy needs of such a farm economically and safely. It is easy to envisage a methane digestion plant becoming one of the common, if not standard, installations on a modern farm. In addition to providing energy and solving the problem of waste disposal, methane digestion has the further advantage of retaining the nutrients in the organic material. The digested sludge is an almost ideal fertilizer. One of the pioneers of methane digestion for farmers is John Fry whose plant in South Africa produced 226 cubic metres per day using the organic 'output' of 1 000 pigs. He wrote:

> I did not attempt to maximize my returns on fuel, fertilizing material or labour saving. It was enough to sit back and enjoy the excess of farm power, the lessened labour requirements, the almost total absence of flies and smells and the independence from outside sources of energy. Also I found my effluent in great demand for playing fields, golf courses, etc., since it promoted growth in the spring, weeks ahead of other fertilizers.[40]

Some commercial firms are now showing an interest in producing modular methane digestion units, evidence that the process is becoming economical. But even with the most optimistic projections the effect on total energy consumption of such developments is small. The delivered energy used by agriculture in the UK, for example, is just over 1 per cent of the total used in the country.

In the developing countries where incomes are low and energy consumption is restricted by cost and scarcity methane production shows considerable promise where social and climatic conditions are favourable. Nowhere is this more striking than in China, where the technology of methane, or 'biogas', production has been refined and disseminated in a remarkable way. The term biogas is coming into general use in recognition of the fact that the gaseous product of the biological digestion of organic matter is in fact a mixture of gases. It is about 60–70 per cent methane, with the rest consisting of carbon dioxide, and some nitrogen, carbon monoxide, hydrogen and hydrogen sulphide. In China, although there have been experiments taking place since the 1950s, large-scale development has only happened in the 1970s. Now there are over 8½ million biogas pits in use in China – most of them for individual households. Using the wastes from humans and animals they provide the cooking and lighting needs of a family, using a digestion pit of 5–10 cubic metres' capacity. Larger communal pits provide gas for use in pumping engines or even for electricity generation.

The rapidity and magnitude of the Chinese achievement is impressive. It is based on meticulous care and attention to design and construction; on frugality and ingenuity in the use of materials – China is a desperately poor country; and on total involvement of all the people concerned with the building and use of the pits. Here is a quotation from the training manual used throughout China:

The pits must be absolutely hermetically sealed so that the whole pit is watertight and the gas sections are airtight. This requires conscientious work and a strict scientific attitude throughout the process of construction . . . any slackening of attention to quality in the building of these pits will interfere with normal gas production, affect the durability of the pits, and may even require far more work to remedy defects . . . So before the pit is built there should be exhaustive study and discussion of its size, the model to be used, the location and the materials.[41]

If the Chinese experience can be made relevant to other countries there is a large potential for the alleviation of hardship in many rural areas of the developing world. China has shown what can, in practice, be achieved with training, motivation and local involvement. It remains to be seen whether other countries will be able to evolve a formula for themselves which will enable them to develop their own biogas programmes. Only India, with 120 000 digesters, has made any significant progress to date.

Biogas technology is about as benign as it is possible for any technology to be. It takes waste materials and extracts from them what is still useful. It provides a further major benefit to rural health in that it kills many of those pathogens in human and animal excreta, which if they make their way into drinking water produce disease. Its waste product is an excellent fertilizer; returned to the soil from which it came it completes the biological cycle. It is not a big technology nor an indiscriminate one; it certainly produces no universal answers to energy problems.

Recently there has been a flood of proposals, calculations, research, and experiments on the production of alcohol by fermentation of plant material. Here the technology is familiar and amenable to economies of scale, and seems to offer the possibility of extracting solar energy from biomass at an industrial level. Breweries, sugar refineries and chemical plants can furnish all the techniques with which this kind of energy harvesting can be developed. It is also attractive to big business since it is one of the few promising means of replacing part of the petrol consumption of conventional vehicles.

Brazil is so far the most advanced. It has an industrial tradition of using the starch from cassava plants and sugar-cane molasses to produce alcohol. In 1976 it embarked upon a large programme of planting sugar-cane and cassava specifically for the production of alcohol to be used as motor fuel. Initially, the alcohol will be added to ordinary petrol up to a maximum of 20 per cent. Beyond that engine modifications are required. Brazil expects to have replaced 20 per cent of its current petrol consumption with alcohol by 1985. After that there are plans for further expansion of production and, possibly, a move towards modifying vehicles so that they can run on pure alcohol.

There are, however, serious questions which remain to be answered. The waste products from fermentation come in large quantities and are extremely pollutive; a litre of alcohol requires the disposal of six litres of acidic waste. There is also a serious question of priorities. In the developed countries the economic return from crops for energy production is unlikely to be competitive with that from food. Hence, the tendency is to look to the developing countries to provide land to grow the energy crops cheaply. Under the pressure of population growth many developing countries may find it difficult to allocate land to energy crops.

It is certainly feasible for countries with large and fertile areas of land to use them to produce plants for alcohol; lakes, too, can be used for growing water hyacinths or other aquatic plants for the same purpose. In most areas of the world, however, there are competing uses for the same land; the choice may be between food and energy. It would be surprising if alcohol fuels were providing more than a fraction of a per cent of world liquid-fuel consumption by the turn of the century.

Pyrolysis and Incineration of Wastes

Pyrolysis is a process in which organic material is heated without air to temperatures of about 500°C. A mixture of liquid and gaseous fuels is produced, leaving behind ash or char. In the United States, where the domestic refuse is of sufficiently 'high quality' to produce about a barrel of oil per ton, large-scale installations have been proposed.

Although such processes are beginning to appear economically justifiable and have their merits as means of saving energy that would otherwise be wasted, they are of limited applicability. Their greatest potential is in societies which are the most wasteful. One of the effects of the rise in energy prices which makes pyrolysis of wastes economically attractive is that it reduces the amount of waste people produce.

Some towns already use an incinerator as a means of refuse disposal. On energy grounds this is almost certain to be more attractive than any other means of disposal and, furthermore, it can be linked to a district heating system. But of course there are snags: many waste products produce noxious fumes on combustion; melting metals can interfere with furnace operation, and optimum operating schedules for an incineration plant do not necessarily coincide with district heating requirements.

While the recovery of energy as gas, oil or simply heat, from waste products has much to recommend it theoretically, there are thus many practical difficulties. In the opulent industrial countries with much to throw out, installations can sometimes be justified for as long as these conditions prevail. In the poorer countries there is so little to meet the direct and pressing needs of their populations that schemes for the recovery of useful energy from their waste products are almost completely irrelevant.

Small-scale Gasification by Partial Combustion

The use of organic materials to produce a combustible gas goes back to the early industrial revolution. During the Second World War, when oil was extremely scarce, small wood-gasifiers were used in Europe to power road vehicles; over 700 000 units were in use at one stage.

The operating principles of a gasifier are extremely simple and the unit contains no moving parts. In essence it consists of a container into which organic material such as pieces of wood, or agricultural wastes like corn cobs, are fed. In the lower part of the container there is a combustion zone, where the organic material is burned. As the material moves downwards in the container, coming closer to the combustion zone, the increasing temperature breaks it down, or pyrolyzes it.

In use the gasification unit is connected to the air intake of a standard, unmodified, diesel engine so that air is drawn downwards through the pyrolyzing material and the combustion zone. The result is that a mixture of combustible gases, including carbon monoxide and hydrogen, is drawn into the engine. This enrichment of the engine's fuel mixture causes the fuel feed control to come into operation and reduce the amount of diesel being used. After a warming-up period, the diesel supply cuts out completely and the engine runs entirely on the gas. If the gas supply is reduced or cut off the engine reverts to running on diesel. The diesel fuel, under normal conditions, is only required for starting and warming up. The saving in diesel fuel can be up to 80 per cent.

Gasification by partial combustion is thus a technology which could see a considerable revival. The possible applications are wide. Wherever there is a sufficient quantity of organic material and a diesel engine there is a possibility of economizing on diesel fuel by means of the gasifier. Although detailed design work would be required to get the gasifier exactly right for different circumstances, and hence no dramatically rapid progress or instant solutions can be promised, this method does seem to offer substantial benefits, particularly in developing countries where petroleum imports are such a burden on the balance of payments.

Part Three

Futures

Thoughts of economy and conservation will inevitably replace those of development and progress, and the hopes of the race will centre in the future of science. So far it has been a fair-weather friend. It has been generally misunderstood as creating the wealth that has followed the application of knowledge. Modern science, however, and its synonym, modern civilization, create nothing, except knowledge. After a hand to mouth period of existence it has come in for and has learned how to *spend* an inheritance it can never hope to restore. The utmost it can aspire to is to become the Chancellor of Nature's Exchequer, and to control for its own ends the immense reserves of energy which are at present in keeping for great cosmical schemes.' – FREDERICK SODDY, *Matter and Energy*, 1912

13

Approaches to Energy Forecasting

Futurology is one of the oldest professions. But from the Pythian priestesses of Delphi to Herman Kahn of the Hudson Institute no one has devised a method of forecasting, with certainty, what is actually going to happen.

The rise of an inspirational leader can transform the attitudes of a country, or a continent; an accident, illness or assassin's bullet can make the difference between peace and war. No one can foretell the breakthrough in understanding which can revolutionize science or create a whole new technology. The rise in oil prices in 1973 took the forecasters and economists of the developed world almost completely by surprise. Major social change is influenced by so many factors that it is almost impossible to predict the shape and attitudes of society twenty or thirty years in advance.

But the future is also unpredictable because so much of it is under human control. A great deal of forecasting goes wrong because it is too mechanistic. It assumes that in the future things must occur as they have done in the past. It ignores the possibility that people will alter their behaviour. The 'doom' literature of recent years has often extrapolated into the future past rates of growth in the consumption of resources or the emission of pollutants and reached quite absurd conclusions. A Victorian, seeing the growth of horse-drawn traffic might have extrapolated a graph of manure deposition in the streets of London and concluded the city would be submerged by now.

The *Limits to Growth* study which caused such a fuss when it was published in 1972 is a good example. It certainly did useful work in focusing attention on a number of important issues and forcing many people to begin to consider the implications of the undeniable fact that the earth is finite. But one serious objection to the study was its clear implication that humanity has little option but to smash itself against the limits to its own expansion. Here is a key quotation:

> Although we have many reservations about the approximations and simplifications in the present world model, it has led us to one conclusion which seems to be justified under all the assumptions we have tested so far. *The basic behaviour mode of the world system is exponential growth of population and capital, followed by collapse* [italics in original].[42]

The study team obviously convinced themselves that the world has no option but to follow the mathematical recipe for disaster which was the basis of their computer model. According to this there must be an exponential rise in population, food consumption, pollution and industrial output matched by a corresponding decline in resources, with the system inevitably collapsing catastrophically some time next century.

This, at least, has the merit of being reasonably definite about things. The book's amazing popularity may have had something to do with its air of pessimistic certainty. But it over-simplifies the real world alarmingly. The world is not a homogeneous entity with evenly distributed resources. Some countries are already in serious trouble, and have been for years; others have sufficient resources to continue in opulence for centuries. Neither is it correct to think of the world as ineluctably committed to exponential growth of the kind shown. The future is not necessarily bound by the laws of the past. It is possible to behave in a different way provided a decision to do so is made. *The Limits to Growth* conclusions demonstrate the valid point that, *if* the world behaves in the way the model supposes it does, and continues to do so for the next hundred years or so, the whole system will collapse. But if, heeding such warnings, humanity decides to behave differently then the results will be different.

The detailed forecast can thus easily become a source of gentle amusement for future generations. When Professor Stanley Jevons was writing just before the outbreak of the First World War, the future of Britain's coal trade was anything but a matter of amusement. Jevons was a perceptive and able analyst, steeped in knowledge of the British coal industry, and concerned, too, with wider economic and social issues. He produced the forecast shown in Table 31 which covers the population, home consumption and coal exports of Britain through until the year 2201.[5] It is a fascinating mixture of hits and misses. Population and energy consumption (measured in tonnes of coal equivalent) for the 1970s are surprisingly close to reality. But he failed to see the importance of oil, completely misjudged exports, and missed the peak in British coal production by 188 years: it occurred in 1913, the very year he was writing, whereas he predicted output would continue to rise until 2101.

It is worth taking the cautionary tale of Professor Jevons further and follow him as he speculates about the time when the world's coal resources are finally exhausted:

> It will not be in the temperate regions of the earth that the great aggregates of population will be situated some four or five hundred years hence, but in the tropics. The population will tend to multiply more rapidly there in the era of peaceful government and with the extension of modern industrial methods . . . there is likely to be, I believe, a progressive concentration of the cruder and coarser manufacturing processes, and also much of the production of bulky goods in tropical regions. As the natives of tropical countries progress under European guidance, and ultimately under their own government, in education, skill and enterprise, they will undertake in their own countries tasks which the more refined Europeans will only do for high wages. When the coal of northern countries is nearing exhaustion, and recourse is had to sun-heat . . . it is the tropics which will have the advantage for manufactures requiring much power.[5]

This beautifully illustrates the perils of extrapolation. Jevons is stimulating when he looks at the implications of a decline in the availability of coal. But he is simply absurd when he begins to spell out the social, political, economic or technical details of life in the

Table 31. 1915 Forecasts of population, home consumption, exports and total output of coal – UK, 1911–2201

Year	Population of Great Britain $\times 10^6$	Consumption per head tons	Annual home consumption tons $\times 10^6$	Exports (including bunkers) tons $\times 10^6$	Total annual output tons $\times 10^6$	Total output to date tons $\times 10^6$
1911	40·83	4·43	180·9	87·1	268·0	900*
21	44·77	4·52	202·5	125·0	327·5	4 175†
31	48·76	4·6	224·0	172·2	396·2	8 137
41	52·73	4·65	245·1	227·1	427·2	12 859
1951	56·62	4·7	266·1	272·0	538·1	18 340
61	60·38	4·75	286·8	314·0	600·8	24 248
71	63·94	4·8	306·9	347·0	653·9	30 787
81	67·25	,,	322·7	375·2	697·9	37 766
91	70·27	,,	337·3	394·8	732·1	45 087
2001	72·96	,,	347·0	411·0	758·0	52 667
11	75·30	,,	361·4	423·0	784·4	60 511
21	77·29	,,	371·0	435·0	806·0	68 571
31	78·93	,,	378·7	443·5	822·2	76 793
41	80·24	,,	385·0	456·0	841·0	85 203
2051	81·26	,,	390·0	468·0	858·0	93 783
61	82·03	,,	393·7	479·5	873·2	102 515
71	82·58	,,	396·4	488·0	884·4	111 359
81	82·95	,,	398·2	497·0	895·2	120 311
91	83·19	4·76	395·0	505·0	901·0	129 321
2101	83·34	4·73	394·0	508·0	902·0	138 341
11	83·49	4·70	392·4	501·5	893·9	146 380
21	83·64	4·68	391·3	488·0	879·3	155 173
31	83·79	4·66	390·5	457·0	847·5	163 648
41	83·94	4·64	389·1	413·0	802·1	171 669
2151	84·09	4·61	387·7	355·0	742·7	179 096
61	84·24	4·58	385·7	294·0	697·7	185 893
71	84·39	4·55	384·0	208·0	592·0	191 813
81	84·50	4·50	380·2	132·0	512·3	196 935
91	84·50	4·45	376·0	72·0	448·0	201 415
2201	84·00	4·40	371·0	38·0	409·0	205 505

Source: H. S. Jevons, *The British Coal Trade*, first edition, 1915; David & Charles, 1972.

*Estimates for 1913 to 1915 inclusive.

†Estimates for 1916 to 1925 inclusive, and so on.

future based upon what he assumes to be the eternal verity of European superior refinement.

The forecaster must therefore remain humble. The main usefulness of his work is that it can reduce the area of uncertainty rather than define the area of certainty. It can demonstrate the implications of certain courses of action and therefore widen the basis for choice. It can also demonstrate that some things cannot happen, or that they cannot happen within a particular span of time. The most useful projections are those which define the constraining limits within which things may occur rather than those which attempt to specify in detail exactly what will happen.

But this can be very valuable. The energy future is too important to be left to wishful thinking or a wistful hope that something will turn up. It is not enough to say that man has fixed things in the past and will do so in the future. He has not fixed everything. The Egyptian, Greek, Roman and other civilizations of the past failed to solve the problem of their own survival. At present many of the world's worst problems remain unsolved. Hundreds of millions more people now endure malnutrition, poverty and disease than in Malthus's time. The US is miraculously advanced in the technology of mass-producing automobiles, but some of its major cities have murder rates which anywhere else would be called a civil war.

Until recently the availability of energy could be taken for granted though, of course, its price and other factors had to be taken into account. Economists debated the merits of coal, gas, oil and nuclear power on the basis of their costs of production and distribution, and the ways in which governmental action through taxes or import controls might affect the competitive balance between them. Energy policy formulation was based on intricately reasoned arguments about the economic competitiveness of coal, oil, natural gas and nuclear- or hydro-generated electricity, and the social and economic consequences of shifts in the production balance between these. It was rare, and considered distinctly eccentric, for anyone to attempt to discuss the implications of the depletion of resources for which no substitute might be obtainable. A continued growth in energy consumption had been a feature of economic development for such a long time that it was presumed it must remain so indefinitely. Investigation of any hypothetical 'post-growth' future was well outside the brief of any economic or physical planner.

But any indefinitely prolonged growth in energy consumption is self-evidently absurd. If world energy consumption were to continue to increase at 5 per cent per annum, as it has done in the past, it would double every fourteen years. Such a process has to end sometime; it is automatically self-limiting. If no other control were imposed, energy consumption would eventually be so high that it would disrupt the biosphere or alter the heat balance of the earth with catastrophic consequences. Energy consumption would then fall because of the disappearance of energy consumers.

As the absurdity of the assumption of indefinitely sustained growth has become obvious, some forecasters have turned to predicting when it must cease. One way of doing this is to draw depletion curves based on estimates of ultimately recoverable reserves. Some of these depletion curves were discussed in Part Two. While they are undoubtedly more realistic than extrapolations of growth into the indefinite future these idealized curves do not take account of the limits imposed on energy production in the real world. In so far as they are based on accurate assessments of ultimately recoverable resources, they usefully outline the upper boundaries of possible future production. Actual production will, in general, be lower since it is influenced by many other factors besides the physical availability of resources.

One such influential factor is the growing appreciation of the dangers of pollution. All but the most irresponsible commercial and governmental enterprises now recognize that preserving the biosphere is not an optional extra: it is a condition of the continued existence of the human race. Oil spills, land dereliction, poisoning of the seas or lakes, contamination of ground-water, ecological damage, disruption of thermal balances: these are no longer regarded as just the whimsical concern of middle-class environmentalists. The dreams of some technological 'realists' that 'the energy crisis will bring environmentalists to their senses' are unlikely to come true. It is too dangerous to ignore the effects of human activity on biological and physical processes of the earth, and a growing number of people realize this. As knowledge increases it is likely that restrictions on the development of energy resources will become more rather than less severe.

Political, social and economic considerations can also influence events. Something like three-quarters of the world's oil reserves are in the Middle East and North Africa. The amount of oil produced in future depends almost entirely on the decisions of the oil producers of these areas. They have a limited capacity to absorb investment and an understandable reluctance to pile up reserves of depreciating currencies. They have a vision of their own future in which some of them will be major industrial powers consuming vast quantities of energy themselves. But, on the other hand, it is essential to them that the world's economic and industrial systems on which they depend do not collapse. They need the technological and industrial output of the developed world for their own development. They also need to preserve a market for their oil. The future levels of oil output will therefore be determined as much by a series of fine political and economic judgements in these countries as by the physical availability of reserves.

The development of substitute energy resources is limited by the rate at which it is possible to introduce new technologies on a large scale. Science and technology can indeed achieve spectacular results in a short time: the Manhattan Project which produced the atom bombs during the early 1940s; putting a man on the moon during the 1960s; or obtaining close-up television pictures of the outer planets during the 1970s. These, however, were just beginnings. Solving the problem of building the prototype is only the first step. The Manhattan Project produced the Bomb, but thirty years' work and many billions of dollars and pounds have not resulted in a nuclear power industry capable of replacing more than a minute proportion of the energy supplied by fossil fuels. And man is still a long way from colonizing the moon or the planets. A breakthrough in the technology of tar-sand or oil-shale development, in super-conductors, or any other area of energy production or distribution, no matter how great its potential, would take many decades to be disseminated sufficiently widely to alter the world energy picture.

It has been suggested by Chauncey Starr that the widespread adoption of a new fuel takes about fifty years.[43] He bases this on the US shifts from wood to coal, and from coal to petroleum for the majority of its energy requirements. In both these cases, the move was towards a more concentrated, convenient and easily available fuel than had been used previously. The task of obtaining energy was simplified each time. The move from petroleum fuels, as they are obtained at present, to any possible substitute is in the opposite direction. It is technically a great deal more difficult to obtain useful amounts of energy from uranium or the sun than it is to obtain the same amounts from oil or gas.

In considering the energy future all these points have to be kept in mind and conventional economic forecasts have always recognized this. They have been based, not on potential levels of production, but on a calculation of the future levels of energy 'demand'.

For an economist 'demand' has a precise meaning. It is the amount of a particular good which people will purchase for a particular price. Thus there is a high 'demand' for good French Impressionist paintings at prices of, say, £500 000 each among the rich art galleries and millionaires of the world. Among ordinary wage and salary earners the 'demand' for these pictures is zero. 'Demand' thus must be clearly distinguished from a vague feeling about what people might like to have. Of course, a large number of people would like an original Renoir in their living room, and at £10 each the 'demand' for these would be very high. At present prices, however, there is no 'demand' for Renoirs among ordinary people.

Even now energy 'demand' is usually calculated by analysing past trends and projecting them forward. Figure 12 which is taken from a report by the Institute of Fuel, in London, is a useful com-

Figure 12. Estimates and extrapolations of world energy consumption.

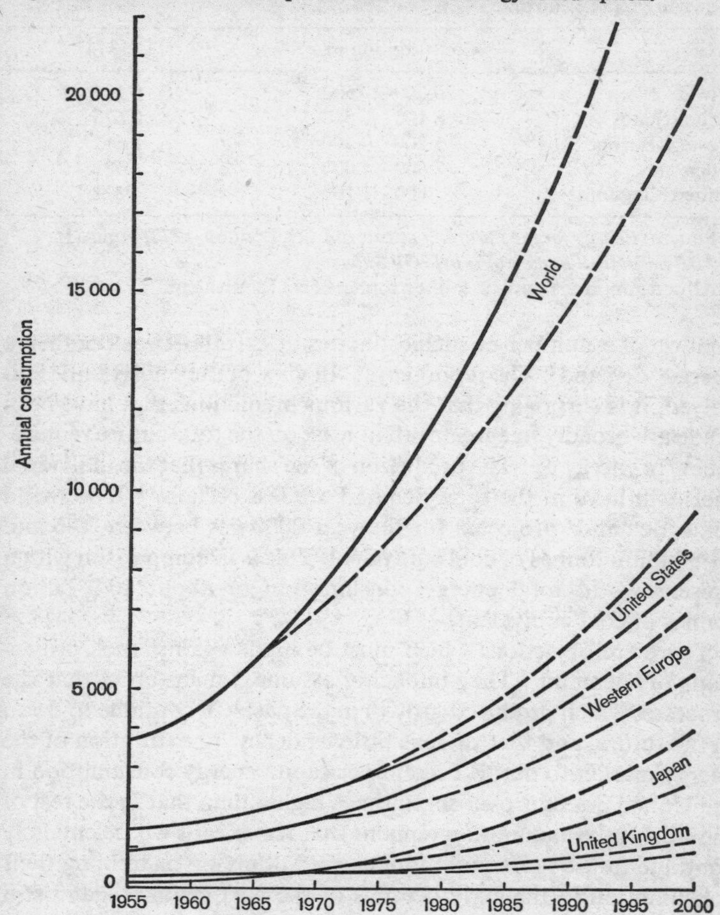

All consumption figures in million tonnes of coal equivalent
(See also summary on p. 254)

Summary of estimates and extrapolations of world energy consumption shown in Figure 12 on p. 253

Area	Consumption in 2000	in 1974*
World	20 000–28 500	9 107
United States	8 100– 9 500	2 675
Western Europe	5 200– 7 200	1 517
Japan	2 600– 4 000	522
United Kingdom	800– 1 100	321

Source: *Energy for the Future*, Institute of Fuel, London, 1973, Figure 1.
**BP Statistical Review of World Oil Industry.*
All consumption figures in million tonnes of coal equivalent.

pilation of a number of such estimates or extrapolations of future energy 'demand'. The report says: 'In view of the uncertainties involved, it is surprising that the various predictions that have been made are broadly in agreement, at least on the total future requirements of energy . . . The prediction of the shares that the individual fuels will have in the total demand are less certain.'[44] The world total 'demand' projected for the year 2000 lies between 20·6 and 28·5 billion tonnes of coal equivalent. This is in comparison with a present world total energy consumption of about 10·0 billion tonnes of coal equivalent.

There are objections which must be made against such projections of 'demand'. They implicitly assume, for instance, that the processes which produced growth in the past will continue to do so in the future, and that there is little tendency for saturation of the energy market to occur. Present per capita energy consumption in the United States is over seven times higher than that in the rest of the world; the assumption remains that Americans will relentlessly continue to use more and more energy, doubling or trebling their consumption in the early decades of the next century – and still show little sign of slowing the growth rate. It is difficult to imagine how they could actually manage to find ways of using so much energy.

In the World Energy Conference of 1978 there were more reservations and qualifications of the energy projections but the basic message was similar: 'World energy demand in the year 2020 is expected to be between three and four times present consumption if average economic growth is similar to that achieved in the past forty to fifty years and there are vigorous and successful measures to improve the efficiencies with which energy is used.'[45]

Table 32 shows some estimates of future populations in the developed and developing regions of the world. It is instructive to compare the data in Table 32 with those of Figure 12. The total projected population increase in the developed regions between 1974 and 2000 is just under 22 per cent. The increase* in energy 'demand'

Table 32. Estimates of world population – developed and developing regions – 1950–2000

Date	Developed regions population (millions)	Developing regions population (millions)	TOTAL (billions)
1950	857	1 649	2·506
1960	976	2 019	2·995
1970	1 084	2 537	3·621
1974*	1 124	2 809	3·933
1980	1 183	3 218	4·401
1990	1 282	4 064	5·346
2000	1 368	5 039	6·407

Source: UN Projections and Estimates, March 1974, quoted in *Food Policy*, November 1975.

*1974 figures interpolated linearly between those in the original table.

taking the US, Western Europe, the UK and Japan as representing the developed regions is 282 per cent; the energy 'demand' is thus anticipated to rise at almost thirteen times the rate of population increase. In the developing regions the population rise is predicted to be 79 per cent and the energy 'demand' increase is just 34 per cent. While per capita energy consumption trebles in the developed countries it falls in the developing countries.

*The mean figures in the range given for the year 2000 in Figure 12 are taken for the purpose of this comparison.

The World Energy Conference of 1978 sees little change in the pattern by 2020. It says: 'The Developing Group now includes 50 per cent of the world's population but uses only 15 per cent of the energy. By the year 2020 the share of the population is expected to be about 65 per cent and the share of energy about 25 per cent.'[45]

It makes very clear the nature of this kind of 'demand' forecast. Most people, presumably, would settle for the comforts of the American way of life, given the chance to do so; but in economic terms the only people 'demanding' to do so are those who have the necessary money and choose to spend it in that way. These 'demand' forecasts or projections are thus estimates of the amount of energy people will be willing and able to buy in the future, provided supplies are available. They have nothing to do with opinions about what kind of distribution of energy consumption would be socially or morally desirable.

One of the greatest dangers is that these ostensibly 'value-free' estimates of economic 'demand' begin to acquire a normative status in the eyes of many people. They look at the projections and assume, not only that these represent the inevitable pattern of the future but that this is also the way things *should* happen. In so far as they then act on this basis, they accept and perpetuate the implicit value-judgements on which the projections were based.

A further danger in a 'demand' forecast is that it can, too easily, be taken as a forecast of supply. The constraints upon energy production are ignored. It is assumed that because the 'demand' is there the industry will somehow manage to cope with the problem of supply, as it has done in the past. There are now few grounds for such complacency.

It can be seen, then, that the conventional forecasts of future energy 'demand' are a long way from being the scientifically objective statements that they are often thought to be. They are more or less arbitrary value-judgements and opinions about the future shape of society, and the use and distribution of energy resources within it. An alteration of any of the assumptions on which they are based would produce a different forecast of 'demand'. Predictions of an 'energy crisis' must be seen in this light. The 'crisis' is postulated on the basis that there will be a gap between the predicted economic 'demand' and the possible level of supply. Because of this, and not because it will be short of energy by most standards, Americans can seriously imagine their country in a state of energy 'crisis', at a level of energy consumption twice that of the countries of the EEC at present. No one, on the other hand, seems to consider that Portugal, for instance, with a per capita energy consumption a third that of the UK or a sixth that of the US, is at present in an acute state of energy crisis. Its poor people are not 'demanding' the affluent standard of the US, or the rich countries of Europe.

It is essential to move away from this kind of forecasting which institutionalizes the patterns of the past and uses them as the basis of normative statements, however disguised, about the future. But finding alternatives is not easy. One possibility which has some attractions is to try to develop the concept of 'need' which lurks inside that of 'demand'. This is certainly much closer to fundamental human concerns and, it might be felt, could be used as a basis for forecasting the essential minimum energy the world will require in future. While people may 'demand' colour television and be little the worse for being without it, they unequivocally 'need' food because they die without it. Unfortunately, once the discussion moves away from the bare necessities of life there is great difficulty in finding agreement on what actually constitutes a 'need'.

It is quite easy to specify the 'needs' of other people, particularly if they are poor. It is done by governments when they fix the level of unemployment relief, welfare services and disability pensions. It is also done when aid is given to countries which have been stricken by famine or other disasters. Tents and essential medical supplies are immediately flown in but no one suggests the relief flights should continue and bring in cosmetics, whisky, hi-fi equipment and large family cars. Poor people, particularly if they are in difficulties, are not considered to 'need' such things. On the other hand people at a high level of affluence can be very emphatic about their own 'needs'.

They 'need' petrol for the car, electricity for the dishwasher, oil for the central heating and gas for the cooker; the peasant in Bangladesh only 'needs' enough dung to cook a handful of rice every day.

It is possible to make a forecast of energy 'needs' based on any assumed set of criteria. But it is difficult to see it having much value beyond that of an abstract exercise. The rich countries have no intention of planning their future on the basis of deliberately bringing their energy consumption down to some level of minimum 'needs'. Neither, on the other hand, will poor countries easily agree that there are different standards of 'need' in various parts of the world – that the American 'needs' a big car, the Frenchman a small one and the Egyptian does not need one at all.

The limitations of all the approaches to energy forecasting discussed in this chapter are obvious. The future is obstinate in its refusal to be trapped in the conceptual frameworks of the present. It may, indeed, be that no satisfactory way of producing a numerical statement of the world's future energy 'needs' or 'demands' will ever be found. Nevertheless, thinking about the future in this way produces some important insights. There are limits to the availability of energy. There are also pressures which are driving the consumption of energy upwards, such as development in the areas which are still poor. Forecasters need to turn more of their attention to defining the true constraints on resources, and to devising the best ways of using whatever resources are likely to be available.

14

Tightening Constraints

For an energy scarcity, or 'crisis', to happen it is not necessary that the total amount of energy available should fall. Scarcity can occur when the amount of energy being produced is rising. A change in the distribution of supplies, for instance, may lead to extreme hardship among those who previously had sufficient. Increased consumption of oil by the producing countries with a corresponding fall in their exports would not reduce the total amount of energy produced, but it could have profoundly harmful effects on the economies of the importing countries.

Alternatively, scarcity can occur because the number of potential consumers, or the amount they want, is increasing faster than the supply. The American energy shortages of the early 1970s happened at a time when total energy consumption was rising rapidly. But, as a result of the consumer boom at the time, people wanted to purchase more gasoline for their cars and electricity for their new air-conditioning systems and domestic appliances than the suppliers were able to provide. In consequence, many people had to go without gasoline, and suffer from power-cuts and 'brown-outs'.

Energy scarcities mean lower availability and almost inevitably higher prices. It is, however, difficult to posit a precise measure of scarcity. As the previous chapter showed, concepts such as 'need' and 'demand' are ambiguous and value-laden and do not provide a satisfactory baseline from which the magnitude of future energy scarcities might be measured. But inability to measure a problem does not make it disappear. It is quite clear that the world is not going to have enough energy to fulfil the aspirations of its people during the coming decades.

This can, perhaps, be looked at in terms of the word 'requirements' – using it in as neutral and value-free a sense as possible. If people have cars and dishwashers, and want to use them, they require energy, and will endeavour to get it. If planners create cities based on freeway systems and high levels of car ownership then people cannot get about them on foot: if such cities are to function a high expenditure of energy on private transport is required. If population growth continues as forecast then an increasing number of people will require houses, consumer goods and food, all of which can only be produced by an expenditure of energy. Throughout the world, humanity's energy requirements, to use a Malthusian phrase, are showing an alarming 'tendency to increase'.

One of the major forces driving this increase is economic growth. If growth in the GNP of an industrialized country ceases its economy runs into trouble. All the industries which provide for growth find themselves short of work. There is no need for extra roads, factories, offices and services; society can get by with what already exists. Building workers, architects, engineers, surveyors and others whose role is that of making provision for growth are left with nothing to do. Secondary effects follow because less money is available for cars, consumer durables, holidays and entertainment and consumption of every kind. Unemployment in occupations depending on these begins to increase. All this happens just because growth has ceased; an actual decline in the GNP has even more severe consequences. A fear economists always have is that a reduction in the rate of growth will quickly spiral downwards into a full-scale recession with a loss of business confidence, falling share prices, reduced capital investment, massive unemployment and a government without the taxation revenue which would enable it to stimulate economic activity or relieve hardship with welfare payments. Some economists talk of the 'steady unemployment' growth rate: that is the growth which is necessary if unemployment is not to increase.

Economic growth – at least in its traditional form – obviously cannot continue indefinitely. But governments which do not promote it find themselves faced with rising unemployment and social distress. If they do not amend their policies they tend to be replaced by election, or by revolution. And traditional economic growth seems to have required increased energy consumption. Throughout the world, governments are, therefore, with varying amounts of success, endeavouring to create the conditions under which their countries' energy consumption will increase – unless measures are taken to ensure that the patterns of the past are not repeated in the future.

The perils of assuming a fixed link between energy growth and GNP growth have already been discussed; there is, in fact, little value in trying to use the past correlation as a means of *predicting* the future. Projecting forward on the basis of the past, however, does bring into focus the magnitude of the energy problem the world is facing; it provides a measure of the changes that will have to be made if economic progress is not to be halted by lack of energy.

If the link between energy growth and economic growth, 'the energy coefficient', is taken to be unity, as it has actually been in many countries, then, with an economic growth rate of 4 per cent per annum, which few planners would consider excessive, energy consumption doubles every 17 years. By the year 2015 the world would be consuming about 40 billion tonnes of coal equivalent. If there is one thing that can be confidently predicted about the energy future, it is that this quantity of energy will not be available by then. The only question is where the shortfalls will be most severely felt and how well the world adjusts to minimizing them and dealing with them.

GNP as a measure of a country's activities, of course, has many weaknesses. It is a crude, highly aggregated index measuring quantity rather than quality. Because of this some people have wondered if it might not be possible to escape from the growth treadmill by adopting a different measure of national activity and redefining the GNP. If its definition were broadened to include the amount of leisure people enjoyed, improvements in the general health of the population, enhancement of the environment and other additions to the quality of life, and were, perhaps, called the gross national welfare, some of the problems caused by the past apparent link between energy consumption and GNP might be resolved. Increasing gross national welfare would not necessarily increase energy consumption. In fact, since GNP includes the cost of such things as road accidents, the chemicals used in clearing up oil spills, the hospital care of people suffering from industrial diseases, and all the energy lost because of inadequate insulation of buildings, it should be possible to increase gross national welfare and decrease energy consumption. With an increase in gross national welfare as their objective, governments – it is argued – would be able to ignore the pressure for crude economic growth, with its concomitant increase in energy consumption, and concentrate instead on making life more pleasant for people. In this way the dual problem would be solved: the electorate would be happy and energy consumption would not be rising.

The world is not so simple. It is doubtful if any government is so crass as to aim solely at increasing its country's GNP, in itself. The primary concerns of most governments, particularly those ruling with popular consent, are that people should have secure employment and that their expectations of increasing material welfare should show some reasonable signs of being met. It so happens that, for all its imperfections as an index, rising GNP usually reflects some success in these objectives. In other words, when GNP is rising people are becoming more prosperous and unemployment is low. No government would dare claim success in its economic policies if GNP were rising and unemployment and poverty were also increasing. The problem of industrial society's dependence on continued economic growth will not be solved by any cosmetic redefinition of terms. Inventing an index of welfare and applying it without changing the structure of the economy would leave the basic problem untouched.

Finding ways of making society function effectively and compassionately in times of economic stagnation or decline is one of humanity's greatest problems. Some economists have begun theoretical investigations of the hypothetical 'steady-state economy' – an economy with zero economic growth. Their objective is to develop a way of applying the methodologies of economics to the activities which would go on in such an economy. To do this, the simple traditional criterion of the maximization of consumer satisfaction has to be abandoned. In the 'steady-state economy', as opposed to the real world, people's desires to consume more would diminish as their consumption increased.

Some of these ideas are interesting; some, implying a freezing of existing social structures, may be pernicious. Most are probably irrelevant as there is little, if any, evidence to suggest that 'steady-state economics' will find voluntary acceptance in practice. People's desires for consumption do not appear to become satiated at any level which might, even remotely, be seen as applicable on a world-wide scale.

A different kind of 'steady-state' is all too easily envisaged. When growth ceases or economic decline occurs in a country its effects are distributed with the casual brutality of a natural disaster. They include unemployment, poverty, business bankruptcies, cuts in social and welfare services, hunger and increasing hardship, particularly for the poor. Until it engages seriously with these problems and that of the present maldistribution of wealth throughout the world, speculation about the 'steady-state economy' will remain an abstract, and academic, economic exercise.

The problem of economic growth is entwined with that of population growth. To exist, people need energy. Some impression of the magnitude of the task facing the world can be obtained by looking at just one of its aspects: the amount of extra energy which will be required for domestic use within the developed countries by the turn of the century. Here is a quotation from the United Nations book on human settlements published after the 1972 Conference on the Human Environment.

The sheer scale of housing need is daunting. It has been estimated that before the end of this century between 1 100 million and 1 400 million new housing units will be needed in the world. Assuming a medium estimate of 1 250 million, an average annual output of more than 40 million new houses is required, about 10 million in the developed countries and 30 million in the developing countries.[46]

The total number of new dwellings required in the developed regions is thus about 200 million by the year 2000. Assume these consume energy at the rate of UK dwellings in 1978, that is about 23 000 kilowatt-hours of delivered energy per year. Assume, to take an extreme case, that all this energy is delivered in the form of electricity. The total required would be $4{\cdot}6 \times 10^6$ gigawatt-hours per year. This is two-thirds of the world total electricity generated annually for all purposes and it would be required just to supply the new houses of the developed world. No one, of course, is suggesting that these would obtain all their energy in the form of electricity; nevertheless the figures illustrate the magnitude of the task implicit in supplying the projected number of dwellings with a reasonable amount of energy from any source. This is an example taken from the rich countries, many of which are in a position to buy the energy they require. In the poor countries which rely on wood or dung for fuel there does not seem to be any way in which their energy supplies can be raised to the level required for projected industrial and domestic use.

The problem of feeding the increasing population of the world is also linked with that of future energy supplies. The eminent economist and writer on demographic topics, Colin Clarke, has produced an elaborate calculation of the world's supportable population which completely ignores the question of energy.[47] He assumed it would be possible to bring all the world's marginal and poor land into cultivation by irrigation and forest and scrub clearing, and to boost yields everywhere by the use of fertilizers, multiple cropping and modern farming techniques. His conclusion was that it would be possible to feed a world population of 47 billions at American standards of nutrition and one of no less than 157 billions at Japanese standards. A simple calculation shows how preposterous these conclusions are.

Assume that UK standards of farming efficiency were reached

throughout the world – a reasonable enough assumption when it is considered that in Colin Clarke's vision of the future a great deal of the farming would have to employ extremely energy-intensive methods. It takes just about half a tonne of coal equivalent to produce a year's food for one person in the UK, which has a rather lower level of food consumption than the US. Moreover, this omits the energy used in food distribution and preparation. Simply producing the food for 47 billion people by these methods, without distributing or cooking it, would therefore require 23 000 million tonnes of coal equivalent per annum, which is nearly three times the world's present annual energy consumption for all purposes. Obtaining this amount of energy every year and channelling it into agriculture is obviously impossible to achieve in any future which can reasonably be foreseen. Indeed the problem of finding enough energy to supply even the world's presently projected population by the end of this century is daunting enough. Anyone attempting to minimize it and induce complacency about the growth of population does the human race an extreme disservice, no matter how high-minded or moral his intentions. Colin Clarke was, incidentally, an adviser to the Pope's commission on birth-control.

These are all problems arising from growth in various forms. It is more disturbing to realize that humanity needs an increasing amount of energy simply to stay in the same place. This would be so even if the problems of stabilizing population and creating a steady-state economy had been solved everywhere. Easily accessible resources of all kinds are becoming scarcer. It is becoming necessary to search longer, dig deeper and use lower grades of ore to obtain the metals and minerals on which society depends. When it is announced that it has become 'economic' to exploit a lower grade of ore this is rarely because it has actually become easier to do so. It is because the richer deposits elsewhere, which were the former basis of supply, have been depleted to a state where they cannot now meet all the world's requirements. The lower-grade deposits must be mined to sustain the level of supply. The same thing occurs with energy itself. As oil and gas from conventional wells become more difficult to obtain the oil industry moves to more energy-expensive sites offshore and in the Arctic, and begins to consider the techniques of extraction from the tar sands and oil shales, or investment in nuclear power.

The pressures driving energy consumption upwards are thus manifold. Luckily, there is still slack within the consumption patterns in the developed countries. These can trim back waste, redeploy work forces and absorb the effects of price rises without collapsing into chaos. The increasing oil prices in recent years, although they have produced difficulties, have not been the catastrophe many would have predicted.

A guess, and it can be no more than that, based on the analysis of energy resources in Part Two, would indicate that the world will be lucky if its total energy consumption increases much more than 50 to 70 per cent by the turn of the century. On past patterns, that is not enough to sustain even moderate economic growth; it is not enough to lift the burden of poverty in the developing world; it is certainly not enough to obviate growing international competition and tension over the allocation of energy resources. The consequences of such a gap between energy supplies and what people are planning to consume will be felt, with varying severity, by everyone.

The days when it was possible to extrapolate a 'demand' curve showing an indiscriminate increase in energy consumption with a confident expectation that the 'demand' would be met are gone for all but a fortunate minority of countries. Inexorably, humanity is entering a future in which the pressing needs of food and shelter will increasingly take precedence over those of convenience, leisure and the pleasures of wanton consumption.

The world, however, is an abstraction in this context. It has no government and no way of allocating its resources except by the mechanisms of trade, and, generally, self-interested aid. No one controls the level of world energy consumption: it is the incidental aggregate of the energy consumption which each individual country has managed to achieve for itself.

It is a fact that charity still begins and, for the most part, ends at home. No country can afford to plan its future on the assumption that the rest of the world, motivated by kindness, will keep it in the style of energy consumption to which it aspires, or feels entitled. Each country must make its own realistic assessment of the amount of energy it is likely to be able to obtain within the tightening constraints on the world energy supply. Having done that it can begin to make plans about how to use it. Its social, physical and economic planning must take place within the budget of its available energy.

15

The Energy Budget

Planning should be a process by which aspirations and resources are matched. It should never produce proposals which cannot be implemented because the resources they require will not be available. It is true that a lot of what goes under the name of planning does result in unfulfillable plans. This is merely drawing pictures of what people, generally the planners themselves, would like to see happen: it would be less confusing if it were called daydreaming and the word 'planning' were reserved for what Hasan Ozbekhan has called 'informed decision and calculated action'.[48]

Without a budget or inventory of available energy, it is impossible to test the feasibility of any proposed course of action which will require an energy investment or which will incur an energy running cost. In producing such an inventory, a country must evaluate the magnitude of its indigenous resources of coal, oil, gas, hydro power, uranium, peat and any other energy sources it may possess. But simple possession of resources is not enough. It is also necessary to calculate the amount of capital the development energy resources will require, how long it will take, what level of production can be obtained, how much money miners and other workers will need to be paid, and how long reserves will last. It is also essential to look beyond indigenous resources to those which will have to be imported and make assessments of how much of these it will be possible to obtain.

A good example of how this might be done is the work of the Government of Canada, in its study entitled, *An Energy Policy for Canada*. In its foreword this says:

Energy, how it is obtained and how used, has emerged as one of the major public questions of our time... Because the sources of energy now employed are known to be finite, and in the face of the increased demands upon them, grave concern has grown up as to whether future supplies will be available, and available at a cost which will not negate our other aspirations... We must soon decide at what rate we are to develop our frontier sources of oil and gas, with all the implications that such development has for those that make their homes in those regions, for the environment of those areas and for the national economy... Decisions now as to how we employ our finite research and development resources will determine the ability to respond to the problems that will affect coming generations... During the time that the studies reported on by this document have been underway, energy reports have been issued by the provinces of British Columbia, Alberta, Ontario and Quebec. The purpose of this report is to define more clearly the national framework into which provincial studies fit, to identify policy choices which must be made within the federal jurisdiction, and to provide a basis for choice by the Government and people of Canada.[26]

It is difficult to see how this can be improved as a brief for a national energy study. It identifies all the major issues and proposes a strategy for studying them which will yield the necessary information in the form in which it is required for choice and decision-making. If it is not quite the way in which energy policy is actually formulated in Canada at present, it remains a sane and pertinent piece of advice.

Difficult though it may be, a country must also make a cold-blooded assessment of its potential supplies from the international market. International bargaining for energy is no different from any other kind of international bargaining. It has to be conducted in the constantly changing scenery of international politics, with its shifting political allegiances, alterations in the balance of power and 'destabilizing initiatives' by major and minor powers. Each country must identify its possible trading partners and make sure it has available for export to them the goods, services or diplomatic friendship for which they will be prepared to exchange their oil, gas or coal. Without this identification of secure sources of energy-supply the credibility of any plans for national development, improving living standards, feeding increased populations, accommodating increased amounts of traffic or increasing GNP must remain in doubt. They belong to the realm of daydreaming rather than that of 'informed decision and calculated action'. In this work of obtaining energy supplies the diplomatic and trading skills of a country's ambassadorial staffs abroad can make as large a contribution as the abilities and efforts of its technologists and coalminers at home.

Orderly planning demands orderly energy supplies; but it is sheer self-delusion to pretend that energy supplies are somehow going to be exempted from the normal perturbations of world trade and politics, the hazards of technological development or the conflicts of industrial relations. The assessment of a country's available energy will continually change. Some programmes for indigenous energy production will prove more difficult to implement than expected, others less so. The level of imports will be affected by international events and by the relative rise and fall of a country's trading strength. There is thus a need for a regular revision of all forecasts of a country's energy availability, probably on an annual basis, but certainly with a major reconsideration as part of the normal five- or seven-year planning cycle adopted by many countries.

All major planning exercises should be obliged to incorporate a means of assimilating and adjusting to revised data on energy availability. Many planners would protest that schemes for, say, highway construction or urban development take five or more years to prepare and could not reasonably be made subject to such chopping and changing. But the alternative is to ignore the importance of the energy question and proceed obstinately with the implementation of projects which have become irrelevant.

The role of international organizations such as the UN and its agencies, OECD, COMECON or the EEC in ensuring that the assumptions about energy availability made by individual countries are reasonable is one which needs to be developed much further. These international bodies are in a position to coordinate and check separate national forecasts against the likely total availability of energy. They may not be in a position to enforce a sharing of energy but they can ensure there is a sharing of information which will enable countries to make their own assessments of their position.

The International Energy Agency's (IEA) annual review of its members' energy programmes and, in particular, their progress in implementing conservation is one initiative which is proving reasonably effective. The subjection of a country's plans for consumption of energy, for substitution of oil by other energy sources, and for conservation, to the inspection and comment of an international panel in the knowledge that the results will be published is salutary. This review system could be the beginning of internationally coordinated planning within a global energy budget. It badly needs to be extended to include the developing countries where energy planning is still more often based on hope than on a realistic appraisal of what is likely to be possible.

The question of future levels of car ownership and use provides an important insight into the kind of problems the world will face but which are only revealed by considering the full energy picture as opposed to what is revealed at a national level.

Most countries in the world are still forecasting large increases in car ownership. A comprehensive survey of forecasts of vehicle ownership and use in the major non-communist countries was sponsored by OECD in 1973.[49] This revealed that planners expected the world's total car fleet to increase from 187 million in 1970 to something over 600 million by 2000. Fuel consumption by cars and commercial vehicles, which was 486 million tonnes in 1970, would thus increase to 1 025 million tonnes in 2000, an extra 539 million tonnes. Despite all that has happened in the meantime, the latest IEA Review[50] reveals that transport plans have scarcely changed. The planned consumption of oil by motor vehicles in 1990 is almost exactly what it was previously projected to be. Allowing for the energy costs of producing the fuel the required increase in annual crude-oil production in the year 2000 is almost exactly 600 million tonnes. This is an increase in world crude-oil production of about 20 per cent. It is most unlikely crude-oil production will ever increase by this amount. It is even less likely that if it does the increase will be devoted entirely to fuelling motor cars. Moreover, since the projected increases in traffic are, for the most part, supposed to occur in the industrialized countries the additional oil will have to be imported. This would require an increase of a third in the amount of oil entering international trade.

Unless there is a massive effort, far beyond anything revealed so far, to shift oil consumption away from its other uses towards transport the world's traffic planners are collectively working on the basis of an untenable assumption about world energy availability. It must be the task of international organizations to bring such anomalies and contradictions into the open. If this is not done the present absurd situation in traffic planning will be repeated indefinitely in other areas of national planning.

Once an estimate of the energy likely to be available to it has been made, a country is in a position to begin to think about how to allocate this between various possible end uses. Proposed developments can be costed in terms of their required expenditures of energy, as happens now with financial costs, and adjusted or rejected if these are too high. It also means that governments must begin to consider the need for systems of energy allocation within sectors of the economy. Hitherto this has only been done, as in the case of petrol or fuel rationing, in times of emergency. But in a world in which many countries are not going to be able to obtain as much energy as they wish, governments will find it increasingly necessary to ensure that the available energy is used to the maximum national advantage – however they may wish to define this.

An increasingly popular way of trying to visualize how this might be done is by the construction of 'scenarios'. These are imagined pictures of the future incorporating a wide range of those distinctive elements which determine the characteristics of a particular society. A 'low-energy scenario' will therefore try to produce a comprehensive picture of a low-energy society in all its aspects.

The Ford Foundation sponsored a major study of the energy choices before the United States, and as a part of this developed three widely differing energy scenarios for the country.[7]

The first scenario was called the 'historical growth' model. This assumed that energy consumption would continue to rise along the same path as it has done historically. This would mean that US energy consumption would increase by about $2\frac{1}{3}$ times by the end of the century. The major implications of this would be that it

> would require very aggressive development of all our possible supplies – oil and gas onshore and offshore, coal, shale, nuclear power. If it proved feasible to increase oil imports on a large scale, then the pressure on domestic resources would relax somewhat. Still, the political, economic and environmental problems of getting that much energy out of the earth would be formidable.

Such a scenario, according to the study, implies that economic growth could continue without interruption and that no allocation of energy or energy-conservation measures on a large scale would be required. But Americans would have to endure the consequences of a maximum development of indigenous resources and a heavy dependence on imports. And if there were any failure in any area of energy resource development the scenario would be impossible to realize.

The second scenario considered was called the 'technical fix'. This 'reflects a determined, conscious national effort to reduce demand for energy through the application of energy-saving technologies'. Energy consumption increases by about 75 per cent by the turn of the century. In this scenario the main feature of society would be 'a market place in which energy is priced to reflect its true costs to society'. It would require innovations like a 'Truth in Energy Law' which would compel manufacturers to provide a label for 'automobiles, appliances and even homes which clearly spells out average energy use and operating costs'. Society would be more energy-conscious but not markedly different from that of today. The pressure to develop energy resources would be much less than in the 'historical growth' scenario, but a major effort would nevertheless be required.

The third scenario was called 'zero energy growth' but was in fact, rather mistitled. It represents an increase in energy consumption of nearly 50 per cent. But

> it represents a real break with our accustomed way of doing things. Yet it does not represent austerity ... It would substitute for the idea that 'more is better' the ethic that 'enough is best' ... Redesign of cities and transportation systems would be a must. Growth in energy-intensive industries like making plastics from petro-chemicals would be de-emphasized.

The 'zero energy growth' scenario thus envisaged a future in which energy consumption rises for a while and then levels out. American society, in fact, was imagined to make the voluntary decision that it had reached the end of growth and that 'enough', indeed, 'is best'.

The study comments on its choice of scenarios:

> Of course, an infinite number of futures is possible; and it is most unlikely that the real energy future of the United States will conform closely to any of the three scenarios we have chosen to describe. They are not predictions, but a tool for rigorous thinking. We do not advocate one option over the others but present each for comparative analysis by the reader.

None of the scenarios was deemed to be infeasible. In fact, they were explicitly described as 'three plausible but very different energy futures'. It is a measure of how quickly and how much the energy picture has changed in the past five years. No one could now take the historical growth or technical fix scenarios as even remotely plausible; even the 'zero growth' looks improbable as prospects for nuclear power and alternative sources of oil fade into the further future.

Accepting the reality of strictly limited or declining energy resources is, however, extremely painful and producing useful scenarios based upon this is difficult. The construction of a detailed scenario for a country is a major exercise. It requires access to a great deal of data; it must be based on realistic professional appraisals of what is and what is not practically possible; and it must take into account the interrelationships of social, political and economic forces within a country. A detailed energy scenario requires the enthusiastic collaboration of representatives of government, industry, trade unions, consumers and energy utilities.

Producing a good scenario is therefore expensive and time-consuming. It also cuts across traditional technical, political, professional and bureaucratic demarcation lines. Even in a country such as Britain where coal, gas and electricity industries are all state-controlled, commercial competitiveness between them is fostered by the government. The jealousies and self-protective attitudes which this induces are even more marked in free-market economies, such as the US, where energy is entirely provided by commercially competing organizations. Since the main purpose of scenario creation is to establish the criteria for choice in a real situation, the more pessimistic the scenario the more defensive the various interests represented become. Many of them will refuse to concede, even as a hypothesis, anything which they feel weakens their position. The Ford Foundation study's preliminary report, for example, contains strongly dissenting memoranda from senior members of the oil, aluminium, nuclear and electrical industries who sat on the advisory board.

With all these difficulties the tempting course of action for any government is to hope things turn out for the best. Few people gain political office through being pessimistic about the future. Asking people to make real sacrifices is bad enough; to be subject to political obloquy as a result of asking people to make hypothetical choices may be too much for the sensitive politician to bear. It would be regrettable if this happened. At its best, scenario creation can be a marvellously useful planning exercise. The scenario which finally passes the tests of practical feasibility and public acceptance is a very secure basis for realistic planning and decision-making.

Energy constraints will not go away if they are ignored. They will still operate. The country which tries to pretend they do not exist and acts without preparing an energy budget will find itself engaged in a continual and futile struggle to obtain more energy than is available at a price it can afford. There will be consequent misallocations of resources and a distortion of economic activity. Any attempt to achieve the impossible diverts resources away from what may be feasible and beneficial. If traffic is not going to double it is pointless; it is, in fact, harmful to construct the roads to accommodate it. In a time of energy scarcity it is better to have well-insulated houses than unused superhighways. If people are going to be jobless and hungry it is better to invest in agriculture and rural development than in grandiose plans for steel-making complexes and new capital cities. If energy is going to be scarce it is better to invest in ways of saving it than in ways of using more of it.

Thus, even if detailed scenario construction is not carried out, there must be an attempt to produce at least an outline energy budget within which planners can work. It is already quite clear that a slow, zero, or even negative growth in energy consumption should be anticipated in many countries. The response to this must be constructive and reasoned. If the dream of doubling car traffic is now seen to be impossible, there is no need to swing to the hysterical counter-assumption that the era of the motor car is over and start making plans for the return of the donkey-cart as the universal means of personal transport. There is still and there will be for a long time ahead a great deal of energy available to those industrialized countries able to pay for it. The task is to recognize their limits and act wisely within them.

But in the underdeveloped countries the outlook is indeed gloomy. Very few of these countries can look forward to ever obtaining the massive amounts of extra energy which would lift them to the standards of living to which most of them aspire. Simple calculation shows the impossibility of raising the anticipated population of 5 billion people in the underdeveloped countries at the turn of the century to American or European levels. Yet, their planners too often persist with the assumption that given time and patience their countries too will reach an American standard of energy consumption.

In the teeming poverty-stricken cities of the underdeveloped world there are embryonic or half-developed motorway networks;

opulent, sprawling, well-serviced suburbs; down-town centres with skyscrapers and multi-storey car-parks; industrial centres with capital-intensive automated industries; and a privileged minority who benefit from all of it. The bulk of the population remains without jobs, services, houses, prospects and, often, enough food to stay alive. The justification given for the disparity of standards is that the affluence will eventually spread to the whole of the population. There is now incontrovertible evidence that it will not. The perniciousness of so much 'aid' given to the underdeveloped countries is that it directs their economic activity into a total dead end. Only in the past few years has the true plight of the rural poor in the developing countries come to be widely recognized. Dealing with the problems being faced there will require a generosity and sensitivity on the part of the developed countries which few have yet revealed.

As the limits of energy availability tighten around all of humanity its range of options becomes restricted. It is imperative that these limits are recognized and delineated as clearly as possible. Otherwise society will engage in the foredoomed and wasteful activity of trying to create a future which is not realizable. But there is not sufficient time available in which to wait for perfect knowledge before acting. The pace of major social change is slow. If society is going to survive within the limits which will be imposed upon it during the coming decades it must begin to make changes now. One of the simplest beginnings it could make is by eliminating waste wherever it occurs.

16

Eliminating Waste – Conservation

Waste could be defined as the consumption of more than is necessary to achieve an objective. It is interesting that in the affluent countries this is even now not necessarily seen as a bad thing. In fact, there are many times when it is advocated as a good thing, and excessive consumption is encouraged for its own sake. Despite widespread speed limits of about 100 kilometres per hour, cars are still manufactured with top speeds of up to 200 kilometres per hour. The capacity for speed in excess of the legal maximum is surplus to all requirements except those of law-breaking, but glib advertising extols its alleged merits and tries to make it socially desirable. Houses and domestic gadgets, too, are frequently portrayed as symbols of superior social status if they are surplus to any obvious needs.

This impulse of modern society to achieve less with more is in marked contrast with the biosphere where any creature which does not make the most of what is available to it is eliminated by more efficient competitors. Wastefulness is an evolutionary dead-end from which no species returns.

The consumer does not benefit from waste. It is a leaking tap, an under-insulated house, lights burning in an empty building, edible food in the refuse pile, or a 300-horsepower car crawling through city traffic. Eliminating waste should be the first priority in any society. It should not require an atmosphere of crisis to justify it. But prosperity has conditioned people into believing that avoiding wastefulness is something to be done only in times of emergency. Viewed with any detachment, such an attitude is absurd. All consumption depletes finite resources and ultimately brings scarcity, but waste brings hard times sooner, makes them worse than they might have been, and ill equips people to deal with them.

Conservation, which is simply another name for cutting down on waste, has many negative connotations in the modern world. It is associated in the minds of politicians and the public with hardship, scarcity and imposed frugality. It seems a regression and an admission of failure. Properly understood, conservation is, of course, none of these. It is an advance, not a retreat. It does not imply a return to primitive technology but an acceptance of the skills of the present. It means producing more with less; it is liberating rather than oppressive; it is the only logical way forward.

Nowhere has energy use reached its theoretical efficiency; in most cases the operating efficiency is but a small fraction of the thermodynamic limit. This means there is a possibility of improvement in every single use of energy. In all the myriads of ways in which energy is used the same result could be achieved with less consumption.

Take for example the petrol-driven car. Higher compression ratios, better ignition control, leaner petrol–air mixtures, improved transmission systems, lighter body-weight, and a host of other minor improvements could halve the petrol consumption of the average European or Japanese family car without impairing its performance. Such improvements could be implemented within the next fifteen years.

In the United States the scope for savings is a great deal more. The petrol consumption of American cars is twice that of those made abroad. There is now bitterly opposed Federal legislation with the long-term aim of reducing the fuel consumption of American cars by half over the next twenty years. If Americans could build cars to the standard of those made elsewhere, that saving would be obtainable much sooner. By halving the consumption of their 110 000 000 gas-guzzlers they would save over 100 million tonnes of oil a year, a quarter of their imports from OPEC, and more than the whole oil consumption of China's 800 million people. It is a saving well worth making since it would go a long way to easing both their own import problems and the tension which is likely to be endemic in world oil markets over the next decades.

That is just the beginning. There is no technical reason why a car consuming 5 litres per 100 kilometres (56 miles per gallon) could not be mass-produced for car markets everywhere. Such cars exist and are already on the market. They are fast, safe,comfortable vehicles. They enable anyone who possesses one to travel with a speed and comfort far beyond the dreams of most of the world's population. If these cars were adopted throughout the world there would be a saving of oil amounting to about 200 million tonnes per year. It could be achieved without the motorists of the world reducing their travel by a single kilometre. If, in addition, all those short journeys to the corner shop, nearby friends or the local park or pub were made on foot, people would suffer no hardship and might be healthier, and the energy savings would be even greater.

This may seem an academic exercise in hypothetical figures; in fact, it is a matter of considerable importance to motorists and motor manufacturers everywhere. Unless such savings are made the market for mass-motoring is not going to be able to expand without cutting into oil consumption elsewhere. Beyond the next decade it will be forced to contract as the availability of oil in the international market begins to decline drastically.

The present level of energy consumption by the world's motorists is an example of short-sighted and self-destructive profligacy. Like a vast herd of goats over-grazing the pasture on which they depend, they are in danger of consuming themselves out of existence. If the motoring organizations, which so vociferously oppose any curtailment of motoring activity, really had their members' long-term interest at heart they would be clamouring now for restrictions on any form of needless vehicle fuel consumption.

One thing they could well ask for is petrol rationing. It is the simplest and fairest way of protecting motorists from their own folly. But even the petrol-rationing schemes which have been used, and are still held in reserve in some countries, seem perversely designed to encourage wastefulness. The ration allocation is usually based on the size of car a person possesses: those with large cars are given a bigger petrol allowance than those with small cars. Any logical rationing scheme would, of course, make a fixed allowance to each motorist. Those responsible enough to use a small car would be encouraged in their responsibility. Those wishing to use a large car to demonstrate their own feelings of superior social or financial status would not be denied that privilege; but they would pay for it by travelling less. The announcement by governments that allowances in any future rationing scheme would be made in this way, rather than on the size of car, should be made now. It would give manufacturers and purchasers the opportunity to prepare for it and gradually wean the motoring public away from their widespread and harmful addiction to energy-greedy vehicles.

Domestic energy consumption in developed countries with climates such as those of northern Europe is between 30 and 40 per cent of total national energy consumption. In the United States domestic energy consumption is also about 30 per cent of the national total and the same is true of Italy. The obvious climatic differences do not appear to have as great an effect as might be expected. In all cases the amount used for heating and cooling buildings is about three-quarters of the total, the rest being used for water heating, cooking, lighting and domestic appliances. The amount of waste is colossal. Most of it is due to poor building design and low standards of insulation.

Heat is lost from a building by conduction through its fabric, by draughts, and by deliberate ventilation. The proportion of the total heat loss caused by each of these varies greatly between buildings. In any attempt to reduce waste all should be looked at carefully.

In the United States many buildings are constructed without any deliberate provision of insulation whatsoever. The Ford Foundation energy study stated that a fifth of the houses of well-to-do families have no insulation, and that this proportion goes up to 50 per cent of the houses of the poor. Insulation standards vary widely in Europe. Among the worst are those of the UK and the Netherlands. An example quoted in a Dutch study[25] reveals how much could be saved by adopting the insulation standards of Sweden. A survey of houses in the Netherlands showed that the average heat loss with an external temperature of −10°C was 12 kilowatts. In other words, maintaining a constant indoor temperature of about 21°C required the operation of the equivalent of twelve single-bar electric fires. In Stockholm the average heating loss from comparable dwellings, with a lower external temperature of −18°C, was less than half this – just 5 kilowatts. And the loss from some especially well-designed apartments built around 1958 in the Ostberga district of Stockholm was only 2·2 kilowatts. In other words these apartments could be kept warm with an external temperature of −18°C using little more than the incidental heat from lights, cooking and the pilot lights from gas appliances.*

The heat loss through the fabric of a building depends on the materials of its construction and the difference between inside and outside temperatures. The insulation characteristics of a building material are described by what is called its U value. This measures the rate at which energy flows through the material when there is a temperature difference of 1°C between inside and outside faces. Table 33 shows some typical U values for various materials and methods of construction – the higher the U value the worse the performance as an insulant. It can be seen that a solid brick wall loses heat three times more quickly than an insulated cavity wall. A single-glazed window loses heat nearly six times as fast as the insulated cavity wall: the all-glazed wall facing the view can therefore be very costly in energy. The advantages of roof insulation are clearly shown in the table: 75mm (3 inches) of glass fibre or similar insulation in the roof space can cut heat losses to a fifth those of an uninsulated roof.

There is little excuse for not building new houses to a standard of

*Pilot lights are surprisingly heavy energy consumers. A large pilot light, such as that on a gas boiler, uses about 2 000 kWh per year, almost 10 per cent of the total annual energy consumption of a UK dwelling.

Table 33. Standard U values for typical elements of building construction

Construction	U value watts/metre²deg C
Solid 105mm brick wall with 16mm dense plaster	3·00
Cavity brick wall with 105mm leaves and 16mm dense plaster	1·50
Cavity wall with 105mm brick outer leaf, 100mm lightweight block inner leaf and 16mm dense plaster	0·96
Ditto but with 13mm polystyrene board in cavity	0·70
Single-glazed windows	5·60
Double-glazed windows	3·20
Pitched tile roof with felt, roof space and 10mm plaster-board ceiling	1·90
Ditto with 25mm glass fibre	0·71
Ditto with 50mm glass fibre	0·51
Ditto with 75mm glass fibre	0·38

Source: 'Building Research Station Digest 108', quoted in *Energy Conservation in the United Kingdom. Achievements, aims and options*, National Economic Development Office, HMSO, 1974.

insulation comparable with, say, that commonly adopted in Scandinavia. The extra financial costs of incorporating insulation in wall cavities and roof spaces, and of double glazing, are small in comparison with the savings they will bring over the lifetime of the building. Nor need there be any fear that the energy costs of extra insulation will be greater than the energy savings. The amount of energy saved over the 60 years or more of a building's life will more than repay the energy investment in any of the usual methods of improving insulation or thermal performance. The lack of action in imposing building standards in many countries is evidence that their governments still have not grasped the fact that the world is facing an energy problem. Here is a quotation from the International Energy Agency's review published in 1979:

In the residential/commercial sector the potential for economically justified savings remains strong. Since last year's review, a number of countries have introduced building codes or are expanding or strengthening existing codes. However, strong building codes stipulating minimum thermal efficiencies for new buildings have yet to be put in place in almost a third of the IEA countries.[50]

Concern is often expressed about the need to provide economic justification for increased insulation in buildings; some atavistic free-market economists even express the belief that the question should be left entirely to market forces. This of course runs completely contrary to that social concern which has led to the creation of a corpus of minimum standards of lighting, sanitation, structural safety and durability in buildings in all the developed countries; such standards are a mark of the seriousness with which a society considers the welfare of its citizens. Energy performance has now joined the list of such concerns.

But altruism, or social conscience, need not be the only driving force for improved standards of building. The costs of improving building standards up to say, the Scandinavian level can now usually be justified in Europe and America by pointing to the savings in energy they would bring the householders. If the cost of *providing* that energy, through a nuclear power plant or a new coalmine, is calculated, energy saving wins, often by a factor of four or five. From the national perspective, in all the industrialized countries, investment in energy saving is far more cost-effective than investment in the equivalent amount of supply. Undoubtedly, the fact that there is not an organized and powerful lobby for energy conservation, whereas there are strong vested interests in energy supply which include manufacturers, utilities and trade unions, contributes to this selective vision on the part of governments when it comes to questions of energy conservation.

Dealing with existing houses is more difficult but much can still be done. Loft insulation is almost invariably possible; draughts can be sealed; pipes can be lagged. Provided it is done carefully and precautions are taken to prevent condensation, internal linings can be applied to solid walls to improve their thermal performance. Double glazing can also be fitted, particularly in living rooms with large windows. The armoury of weapons which can be used in the fight against waste is extensive and readily available.

The scope for saving in the remainder of domestic energy consumption is less than that in space heating, simply because less energy is used. Water heating tends to use about 5–10 per cent of the total energy consumed in a dwelling, but simple and substantial economies can be made. Showers are much more economical than baths; spray taps for hand washing can also save considerable amounts. In favourable circumstances small solar collectors can provide most of the hot water required during half the year and some preheating during the rest of the year even in fairly high latitudes. They can thus cut the total energy consumption in an average dwelling by perhaps 3–4 per cent but this could be increased by skilful design of the installation and a willingness by users to accept something less than instant hot water on tap at all times of the year. The largest savings, however, can be made by simply using hot water less carelessly and making sure tanks and pipes are well lagged.

The simplest method of all for domestic energy saving is to lower the internal temperature of houses. Since the outflow of heat is proportional to the temperature difference between inside and outside, lowering the internal temperature, by reducing the difference, also reduces the outflow. Many households operate their central-heating systems at up to 23°C (73·4°F), whereas most healthy people, wearing a reasonable amount of clothing, can be quite comfortable, even sitting, at a temperature of 19°C (66·2°F). Lowering the temperature from 23°C to 19°C would save about 30 per cent of the annual heat loss through the fabric of a house in the UK. Turning off the heat in bedrooms, hallways and little-used areas of the house can also yield large savings at little cost in comfort.

The proportion of a country's primary energy consumed by offices, shops, hotels and public buildings is relatively small. In the UK, for example, it is about 11 per cent of the total, but the savings which can be made are quite large. Le Corbusier wrote of the house as a machine for living in, and modern architecture has assiduously explored the ways in which nature may be controlled, manipulated and defied by buildings. The result is a large number of buildings completely unsuited to any era of energy scarcity. None is worse than the tall glass-walled office block. Because of the very low insulation value of glass, a glass-walled building rapidly loses heat whenever the external temperature drops. But when the sun shines brightly it acts like a solar collector and rapidly gains heat, to the extreme discomfort of those inside it. Such a building requires a heavy expenditure of energy on running its heating and cooling equipment if it is to be usable at any time of the year.

The modern approach to lighting, too, has led to hosts of unnecessary problems. In the past, public-health legislation generally ensured that occupied rooms in all buildings had adequate daylighting. But modern buildings are frequently exempted from such legislation and are designed with wide internal spaces which daylight cannot reach. Despite its glass walls a modern office block may have to be artifically lit the whole year round. Since even good fluorescent fittings emit 80 per cent of the energy they consume in the form of heat* this greatly exacerbates the problem of summer cooling.

The modern office block looks the same in Kuwait or Killarney, but the traditional indigenous building in any country is built as a direct response to the climatic conditions of its own area. There is no such thing as the universal traditional building; forms vary on the windward and lee sides of the same hill. They use thick walls, thatch, ventilated corridors, the 'air scoops' of the Middle East, courtyards, overhangs, and cunningly designed and positioned windows and doors. All of these serve to damp down the extremes of heat and cold and hence reduce the problems of heating and cooling. These buildings allow their inhabitants to make the best use of the energy available to them.

If architecture is to be of any relevance to the future it will have

*Tungsten-filament lights are much worse. They emit 95 per cent of the energy they consume in the form of heat.

to throw out the rule books it has written for itself in recent decades and begin again to try to understand how to design with, instead of against, nature. The effective modern building should, like the traditional building, make the most of the natural energies of the sun and the wind for heating and cooling; its fabric should be designed to minimize unwanted gains and losses of heat; and its services should be used only to bridge the gap between what can be obtained naturally and what is required for the reasonable comfort of its inhabitants. This is not advocating that the designer should have windmills, solar collectors and such devices bristling from the building, but simply that he avoids creating problems needlessly. The highly serviced building should be seen as a defeat for designers rather than a triumph of technology.

There is thus little problem in improving upon the abysmal energy-consumption standards of the past as far as commercial and institutional buildings are concerned. Moreover, it is now becoming evident that designing to higher standards of energy conservation is not the financial burden it has often been assumed to be. The building which because of its fabric design neither loses heat excessively in winter nor gains it excessively in summer requires less heating and cooling equipment. Space is saved for productive uses; fuel is saved in running costs, and there is often a saving in the capital cost of the building. There are now many well-documented cases which show that energy-efficient commercial buildings can be cheaper to build than the highly serviced and inefficient buildings of recent decades.

Much can also be done to improve the energy performance of existing buildings. The U K government's Property Services Agency after a survey in 1972 installed heating-control systems in 300 buildings and obtained savings of between 30 and 50 per cent in heating costs. The systems cost £408 000 to install and the savings, during the first year of operation, were £262 000, at 1972–3 prices. Since then rising energy prices and technical developments have even further widened the scope for energy savings. Schemes for the re-use of waste heat from lights, for automatic controls of heating and lighting, for more efficient boilers and heating systems are increasingly available to the users of buildings. Micro-processors, with their ability to handle complex instructions, open the way to highly sophisticated methods of controlling the levels of heating and cooling to match varying needs in different parts of buildings. The device called the 'optimum start control' enables the automatic switch-on of heating in buildings such as schools or offices to be controlled automatically in accordance with the weather; on warm mornings the start is delayed. Energy is saved and the occupants benefit by not arriving at an over-heated building. There are coating films which can be applied to windows to reduce the amount of solar energy absorbed in the summer and hence cut down cooling bills; another kind of film cuts down the rate of radiation of infrared energy from inside the building and reduces heating bills.

Industry typically consumes about 40 per cent of the energy used in industrial countries though this goes up to 58 per cent in the case of Japan. The potential for energy saving is even greater than in other sectors of energy use.

Industrial buildings are notoriously badly designed for energy conservation. Many manufacturing processes are carried out in large thin-walled sheds with open doors and no attempt whatsoever to economize in energy use. Indeed, space and water heating, in much of light industry, can consume more energy than the industrial processes themselves. In the UK it accounts for 36 per cent of the energy consumed in the whole of the engineering sector of industry. This sector accounts for almost half the output of British industry.

Insulation of factory buildings, provision of doors that can be easily closed, control of ventilation and heating systems, and general attention to the energy economy of buildings can sometimes eliminate the need for space heating completely. There are many other effective measures available. Take the question of industrial waste heat. A high proportion of this could be recovered and used, for example to preheat boiler water, or for space heating. The principles of heat recovery are well known and are widely applied in some industries which use a great deal of energy, notably the chemicals industry. The technology of heat recovery is quite undemanding. It usually has to deal with relatively low temperatures, rarely more than 100°C, and performance requirements can usually be met by readily available materials and conventional methods of manufacture. The engineering demands of energy saving are much less advanced than those of energy provision, most notably in the nuclear industry.

One of the most simple and ingenious heat recovery devices is the 'heat wheel'. It is widely used in Sweden where it was invented and developed, but it is not as well known and used elsewhere as it deserves to be. It consists of a wheel, perhaps a metre in diameter and thick in section, made of wire mesh or some equivalent material. If warm air is blown through the mesh, heat is absorbed; if cold air is then blown through it will be warmed by the heat previously absorbed by the mesh. In practice the wheel is installed so that it intersects the inlet and outlet ducts of a ventilation system where, by rotating slowly, it absorbs heat from the outgoing hot air and transfers it to the incoming cold air. The efficiency of heat recovery may be as high as 80 per cent.

Although heat wheels, and other heat recovery systems, can often be installed for less than the savings they would bring in a single year, they are rarely used. Partly, the reason for this is ignorance. Energy has never been a major part of the costs of manufacturing in most industries; or at least it has never been thought to be one on which significant savings could be made without much difficulty. As long as energy was cheap and its supply was assured there was little pressure to give it much attention. Only in those industries where energy expenditure is heavy and processes are complicated, as in chemicals and petroleum refining, has engineering expertise been deployed to reduce losses and recover waste heat. In Sweden, however, where keeping buildings warm is a major problem and can be cripplingly expensive unless it is done efficiently, heat recovery is widely used. Elsewhere, the typical manufacturing enterprise has continued in complete ignorance of its own energy waste or how easily it might be reduced.

Industrial processes can also generally be improved with surprisingly small efforts. Insulation of pipes, repair of leaks, replacement of obsolete or worn-out boilers, use of waste products as fuel, better maintenance, improved lubrication, reduced unloaded running of machinery (which can consume nearly half as much energy as loaded running): the list is almost endless and can be added to by casual inspection of any factory in operation. It is now fairly widely agreed that most industry can show savings of at least 30 per cent over the next couple of decades; all that is needed is the willingness to take advantage of the opportunities which already exist.

No survey of energy wastage can afford to ignore the electricity industry, the largest single consumer of energy in most of the industrialized countries. To make electricity from fossil fuels about two-thirds of the heat content must be sacrificed as waste heat. This is not a condemnation of electricity generation in general nor does it deny the convenience, usefulness and importance of electricity. It is, rather, to emphasize the energy advantages of 'thermodynamic matching' of energy supplies to their end-uses.

Poor-quality coal or residual fuel oil are ideal power station fuels because they cannot be used elsewhere. The same is true of nuclear and hydro power: using them to generate electricity is the only way they can be made to contribute to energy supplies. On the other hand, the use of natural gas for electricity generation, as is still done on a large scale in the US, is a scandal. Natural gas is an ideal fuel for many applications. To waste three-quarters of it making electricity and then to use the electricity in an electric cooker is ridiculous.

Where it is possible to use a fossil fuel directly for heating, rather than using electricity, then there will almost inevitably be a large energy saving. The long-term policy should be to reach a state where no fuel which could be more efficiently used elsewhere is used in power stations.

The development of ways of using the vast quantities of waste heat from power stations seems an obvious objective in any waste-elimination campaign. Unfortunately the waste heat is rejected at such a low temperature that it is difficult to find a use for it. Thus a power station using cooling water will produce gigantic quantities of water at a temperature of about 30°C. Little can be done with this. Even if it were used to heat greenhouses, or for fish-farming, there is an obvious limit on these enterprises.

There is, however, a considerable unrealized potential in power stations deliberately designed to produce both heat and electric power – they are often called combined heat and power, or CHP, stations. In these, a sacrifice is made of some of the generating capacity in the steam and it is drawn off at a usefully high temperature. The efficiency of electricity generation falls to perhaps 20 per cent; but the combined efficiency of the system may be as high as 70 per cent, twice as high as that of a good modern station.

Of course there are difficulties. A CHP station requires an appropriate and reasonably stable balance between the demands for heat and electricity. When the balance changes there may be an excess of heat, which is wasteful; alternatively there may be an excess of electricity which has to be sold to the central generating authorities, or a deficiency which has to be made up by them. The generating authorities tend to dislike such arrangements and are rarely encouraging. Thus, while CHP systems are fairly common in industry, where the balance between the heat and electricity loads is predictable and controllable, they have not found wide application elsewhere. The exception is in Denmark where CHP systems provide 6·5 per cent of total consumption of energy. It is noteworthy that in Denmark, where the local authorities have a high degree of control over electricity supplies, the difficulties which central electricity generating authorities find insurmountable elsewhere have not inhibited the development of CHP.

Another way of making better use of electricity is by the use of a heat-pump. This machine performs the apparently magical feat of taking energy from a low-temperature source, such as a river or the atmosphere, and raising its temperature to a useful level. The 'coefficient of performance' is used to measure the amount of useful energy produced by a heat-pump in comparison with that used to run it. In practical operation, using the atmosphere as a heat source and delivering heat at a temperature of 50°C, the coefficient of performance of an electric heat-pump would be 2·0–2·5. This means that a heat-pump consuming 1 kilowatt would be producing 2 to 2·5 kilowatts of useful heat. The efficiency of a heat-pump declines the greater the temperature over which it is required to 'pump', and preferably this difference should not exceed 50°C. If any low-grade heat source, be it power station waste heat or warm water from a solar collector, can be used as the source for the heat-pump, its performance can be increased.

In the US heat-pumps are familiar objects under another guise: they are used for summer cooling of buildings. The domestic air-conditioning unit takes heat from inside the house and dumps it, at a higher temperature, outside. In a refrigerator the same principle is used. Heat is extracted from the cabinet and given off at a higher temperature from the condenser coil. It is sometimes said that a heat-pump is a refrigerator in reverse; this is not so. It uses exactly the same principle as the refrigerator. Up to now heat-pumps, for heating, have not been economically attractive in the face of competition with other energy sources because of their relatively high capital and installation costs. The economic prospects are however much brighter now and it is probable that industrial and domestic heat-pumps will increasingly be used for heating.

The catalogue of easily avoidable waste in industrial society can be continued a long way. Perhaps it is not surprising there is so much slack to be taken up. Over the past few decades little attention has been paid to energy conservation. The glamorous work for engineers and technicians has lain in the big supply technologies; in designing power stations and transmission systems; above all in nuclear power. This is where a great deal of the engineering and scientific talent has gone. All the millions of end-uses have been left to evolve under the pressures of ephemeral styles, advertising, ease of production and saleability in a market which has been almost totally uninformed about energy and indifferent to it. As long as the pumps, pipes and wires delivered energy at a price most people could afford no one seriously questioned the way they used their energy supplies. No one told them they could have achieved the same result with much less; that their energy bills could have been halved. No one pointed out that the energy policies which were being carried out all over the developed world, and being exported to the developing world, were leading towards the present energy predicament. Demand was taken to be what it was and inevitably so; the forecasters saw it getting larger; the engineers worked to meet it. It seemed to be the natural order of things. Only under the shocks of the 1970s has it begun to dawn on people that things could be different; that, indeed, if industrial society is to survive even the next few decades they must change.

One of the most comprehensive analyses yet carried out on the potential for energy saving in an industrial economy was published in the UK in 1979. It was called *A Low Energy Strategy for the United Kingdom*.[51] It is worth examining in detail because much of what it found is undoubtedly relevant in other industrial countries. It considered two hypotheses about future economic growth: a high case in which the country's total GDP was assumed to treble by the year 2025, and a low case in which it doubled by the same date. The economy was divided into four energy-consumption sectors: industry, domestic, transport and commercial. Within each sector an assessment was made of the energy-conserving practices already in use and capable of being disseminated further and those considered likely to be economically and technically feasible within a short time. These were used as a basis for estimating the energy-saving potential in each of the sectors. It was assumed that these savings would be gradually realized by the implementation of the energy-conservation measures over the next 30 years. No especially vigorous conservation campaigns or government actions were assumed. Neither was there any fall in living standards; quite the contrary:

> ... houses become warmer so that everyone enjoys the amenities of the better-off today. Most families come to own freezers, dishwashers,

clothes driers, colour TV, and other heavy users of electricity. Car ownership grows rapidly so that 72–75 per cent of households in 2 000 have at least one car compared with 58 per cent today . . . Air traffic grows 2·4–3·0 times . . . total industrial output increases by a factor of 1·7–2·2. The areas of offices, schools, hospitals, shops and restaurants increase by anything from 30–80 per cent.[51]

The results were startling and dramatically demonstrate the potential that exists for conservation. In the high case, where GDP trebled, energy consumption did not increase at all. In the low case, where GDP doubled, energy consumption fell by 14 per cent. These are results far below the official forecasts of the amount of energy needed to provide for the same levels of economic growth. If economic growth were to be less than assumed then, provided the conservation measures were applied, the amount of energy required could fall even further. In its conclusion the study states:

> The emphasis on conservation would create a great diversity of jobs, unskilled as well as skilled, in thousands of factories and workshops across the country – in sharp contrast to the specialized, centralized and limited job opportunities implied by the conventional supply-expansion energy forecasts and policies. By reducing the drain on world fuel resources, Britain would in a small way lift one of the heaviest burdens preventing the development of the poor world. If other major energy-using countries followed a similar path, the effects on global development and confidence in the future would be extraordinary.

This study provides a welcome gleam of light in an otherwise gloomy energy future. Conservation, in all its aspects, offers the world an attractive prize. Using the brains and efforts of engineers and scientists, of architects and industrial designers, to grasp it could introduce a time of endeavour as exciting as that of the 1960s' space efforts. It would not be a restrictive time; it would be one in which the pressure to find ever growing quantities of energy were eased and the problems of transition into a future without oil were understood and solved.

One of the major attractions of a concerted effort to eliminate waste is that it could begin immediately. It requires no more than a consciousness of waste, as waste, and a willingness to do something about it. If industrial society fails to seize the opportunity it will not be because the technical resources are lacking; it will be a failure of vision and purpose for which the penalty is likely to be dire indeed.

17

Planning for Transition

In January 1972 the *Ecologist* magazine produced a document called 'A Blueprint for Survival'. This was supported by the signatures of thirty-three prominent academics among whom were Sir Julian Huxley, Sir Frank Fraser Darling, Peter Scott, and seventeen university professors; five of the signatories were Fellows of the Royal Society. A letter written to 'welcome the document as a major contribution to current debate' and signed by another fifty scientists was published in *The Times*. The Blueprint attracted immense national and international attention. One commentator wrote: 'The Blueprint for me is nightmarishly convincing. For those who like myself have regarded environmental considerations as a respectable, but not particularly arresting, type of cosmetic surgery on Industrial Society, it is mind-blowing. After reading it nothing seems quite the same again.'[52]

The Blueprint, and the response to it, must be seen not as isolated events, but as the culmination of a number of trends of rising public concern about pollution and resource depletion. Rachel Carson's *Silent Spring*[53] published a decade earlier in the United States had played a major part in awakening the awareness of Americans, and others, to the dangers of carelessly applied technology. The writings of the Erlichs, Barry Commoner and others during the 1960s had heightened such anxieties; and Hubbert in *Resources and Man*,[17] published in 1968, had predicted a peak in world oil production in the 1990s. The Blueprint, therefore, brought together in a succinct form many of the worries of the preceding years. It took a wide view of the perils the world was facing: energy scarcity was just one of the elements in its diagnosis of the plight of humanity. It differed from its antecedents in proposing a comprehensive solution.

This solution was based upon a detailed programme of population control, decentralization of industry, de-industrialization, de-urbanization, revision of energy-intensive practices in agriculture, and the development of non-depleting energy sources. It called for a comprehensive and 'orchestrated' change in the whole style and structure of society which would lead, over a period of about a hundred years, to the 'stable society' which was defined as 'one that, to all intents and purposes, can be sustained indefinitely while giving optimum satisfaction to all its members'.[20] The Blueprint was concerned primarily with the replacement of industrial society; but the proposed 'stable society' would, presumably, have been an equally acceptable paradigm for the development of the poor countries also.

The objective of a 'stable society' in which all its members obtained optimum satisfaction is admirable; no reasonable person could object to such a happy resolution of humanity's problems. The Blueprint, however, suffers from the fatal weakness of most utopian tracts, from Plato's *Republic* onwards. It is an idea rather than a practical programme for action – despite its protestation to the contrary, and detailed specifications for change. It does not accept the exigencies of place, politics, and people. It presumes to be able to define, in the abstract, an ideal society and then work towards it in practice.

There are a number of dangers in this. One is that pursuit of the unattainable ideal easily distracts attention from what can be done. The huge investment programmes of state and industry are still grinding along, relentlessly foreclosing the options of society. A desire to have nothing to do with them is not, perhaps, the most effective base from which to start influencing them.

But, more importantly, the whole idea of a comprehensive plan for society may be questioned. The history of such ideas is that they are completely ignored or an attempt, generally bloody, is made to implement them, with results which the original thinkers would not have supported. The danger does exist that under the pressure of energy or other constraints an attempt will be made somewhere to implement the idea of the 'stable society', with dispersion of population, decentralization of industry and other recommendations of the Blueprint. But such a 'stable society' would be far from the ideal envisaged: it would be controlled, repressive and restrictive. Idealism and oppression are separated by far too thin a line for most people's comfort.

What is required is not a comprehensive plan, but rather a willingness to accept the need for a complicated organic response to resource and other constraints. The objective must be to heighten public awareness and develop a widespread understanding of the problems facing society – and of the limits within which they must be solved. Without this, action will too easily be misdirected and resources misallocated.

A considerable number who would agree with much of the Blueprint's basic diagnosis of the problem would, therefore, not be happy with its somewhat overweening presumption to specify in such detail what society should do. They would, however, maintain that the case for abandoning the heavily energy-consumptive, centralized, pollutive, urbanized activity of modern society is irresistible. This inspired what came to be widely known as the 'alternative technology' movement. Other names used to describe it include 'soft technology', 'low-impact technology' and 'eco-technology'. In the words of one commentator, David Dickson: 'The approach of each group usually contains some combination of a set of common elements. These include the minimum use of non-renewable resources, minimum environmental interference, regional or sub-regional self-sufficiency.'[54] The movement concerned itself, at a practical level, with the development of small energy-capturing devices: solar heaters, windmills, waterwheels, methane digesters for domestic houses or small communities, and methods of growing food in urban locations. The emphasis in all of this was on home, or craft, methods of operation rather than factory production. The objective was to free people from the domination of large-scale centralized systems. The ultimate extension of this is the 'autonomous eco-house', a totally self-contained dwelling relying entirely on 'ambient' energy sources and recycling all its own waste products.

The results of this activity have been, almost uniformly, disappointing. Many of the reasons for this will be clear from the survey of energy resources in Part Two of this book. Apart, possibly, from solar collectors the energy output of 'alternative technology' devices is negligible in comparison with that from conventional sources and seems destined to remain so. The 'autonomous house' can only be achieved, at enormous expense and inconvenience, by the application of very elaborate technology, and then only in favourable climatic conditions.

It has been, of course, argued that the very point of 'alternative technology' is its irrelevance to industrial society. Here is a quotation from Godfrey Boyle's book *Living on the Sun*: 'The inexhaustible energies inherent in sunlight, in wind and water, in plants, in the earth's geothermal heat, and in the ebb and flow of the tides, provide a far firmer foundation for liberty, equality and fraternity among Mankind than the energy *products* purveyed by the present oligopolistic cartels, both capitalist and state-capitalist.'[55] What concerns this writer is not so much the technology, in itself, as the kind of society it supports.

The same idea comes up in *Alternative Technology* where David Dickson says that 'from the political perspective, any discussion of alternative technology must be necessarily utopian . . . and the various tools, machines, and techniques described will be given the label "utopian technology".' He says of 'utopian technology': 'It forms a framework that is designed to eliminate the alienation and exploitation of the individual, and the domination of the environment by the activities of man.'[54]

In both these authors the utopian vision of society is prior to, and more important than, that of technology. They are idealists with ideas about how society should function rather than technologists. The technology they discuss is hypothetical and, essentially, derived from their concept of an ideal society. Certainly, such a technology, if it did exist, would fit well with their social theory; but it is not necessary for the theory that it should exist. In fact, as the past few years' work has shown, the postulated technology does not exist, and it is noticeable that the emphasis in the 'alternative technology' movement is shifting away from the earlier technological naïveties towards a deeper consideration of the social ills of alienation, and materialist or organizational oppression which it diagnoses in contemporary industrial society. Its prescriptions, with their emphasis on personal freedom and responsibility, tend to have much in common with those of anarchist theorists such as Kropotkin. This area of speculation, in fact, has always been its primary concern; its preoccupation with technology has, paradoxically, been secondary.

It is necessary to understand this. The still wide popularity of 'alternative technology' owes much to the belief that it offers a way of dropping out of the rat-race of industrial society in a relatively painless way. Many people believe, and with 'alternative technology' devices moving into the mass do-it-yourself market are led to believe, that, with the appropriate gadgets fixed to their dwellings, they can free themselves from dependency on the energy supply utilities, and make a useful contribution to conserving the earth's diminishing energy resources. The aspirations are understandable and laudable, but they are bound to be disappointed. The potential contribution of 'alternative technology' to sustaining life in cities such as London, Paris, New York, Stockholm or Moscow is negligible.

Nevertheless, the dream of a more free and decentralized society persists. Nuclear power remains the focus of a great deal of passionate opposition, not just for itself, but also as a symbol of a technocratic and threatening society. The fear is that a society based on large, centralized technologies would impose an authoritarian rule on its people. It is interesting to see how well Aldous Huxley, writing in 1946 and long before any extensive work had been carried out on civilian nuclear power, expressed this present-day fear:

. . . it may be assumed that nuclear energy will be harnessed to industrial uses. The result, pretty obviously, will be a series of economic and social changes unprecedented in rapidity and completeness. All the existing patterns of human life will be disrupted and new patterns will have to be improvised to conform with the non-human fact of atomic power. Procrustes in modern dress, the nuclear scientist will prepare the bed on which mankind must lie; and if mankind doesn't fit – well, that will be just too bad for mankind . . . It is probable that all the world's governments will be more or less completely totalitarian even before the harnessing of atomic energy; that they will be totalitarian during and after the harnessing seems almost certain.[56]

In his book *Soft Energy Paths* Amory Lovins distinguishes between what he calls the 'hard' and the 'soft' path to the energy future. The 'hard' path relies on nuclear power, coal, large-scale developments of shale oil, solar thermal power stations, and all the panoply of complicated energy technologies now being discussed and developed in research laboratories in the US and other rich countries. It is based on the belief that energy demand must continue to grow at its historic rates and that science is capable of developing the means of providing all the energy required. It is a nightmare society and one in which Huxley's fears could well be realized.

In contrast with this Lovins places a tantalizing vision of a society supporting itself on small-scale, benign, renewable and non-polluting energy sources. The route to this he calls the 'soft path'. At the end of it lie:

. . . jobs for the unemployed, capital for business people, environmental protection for conservationists, enhanced national security for the military, opportunities for small business to innovate and for big business to recycle itself, exciting technologies for the secular, a rebirth of spiritual values for the religious, traditional values for the old, radical reforms for the young, world order and equity for globalists, energy independence for isolationists, civil rights for liberals, states' rights for conservatives.[57]

Alas, the case Lovins makes for the feasibility of the soft path is less than fully convincing. His book, crammed though it is with calculations, turns out to be more a passionate polemic for a social vision than a scientific analysis. His excoriation of the mindless technological optimism of those who see large-scale technology solving all the world's energy problems is richly justified. But his own optimism about the potential of small-scale alternatives, and above all about the social changes which would be required to implement them, is not supported by dispassionate observation of the contemporary world. There is little evidence that people are yet content with 'enough' or that social responsibility and concern are beginning to take precedence over individual acquisitiveness, that small is beginning to appear truly beautiful to people and the politicians who make decisions on their behalf.

The transition to a time when oil is restricted to a much smaller range of uses than today is inevitable; but it will be difficult to manage. No one, in fact, knows how to shift from the present dependence upon oil without disrupting the working of industrial society.

Until quite recently it was often, and naïvely, assumed that nuclear power would provide a substitute for oil, and nuclear programmes were announced with that purpose. But nuclear power can only produce electricity, and that cannot be used as a substitute for many present uses of oil. The fact that electricity is available to the consumer at the flick of a switch conceals the elaborate, cumbersome, expensive, and frequently inflexible methods by which it is produced and distributed. A nuclear power station must produce its electricity at a virtually constant rate, day and night; it is connected to the consumer by a system of transmission lines, transformers, switches, and fuses. A petrol pump of the kind found in most filling stations delivers energy at the rate of 30 megawatts and can be turned on, off, or varied at will; a filling station with 20 pumps has a possible energy delivery rate of 600 megawatts. The capital costs of the transition away from oil, and the rigidities it will introduce, are just beginning to be recognized.

The problems which would be faced in a 'soft' path of substitution are probably even greater. Difficulties in the economic system caused by scarcities of oil do not automatically create the market conditions under which 'soft' technologies can easily or satisfactorily be introduced. Take, for example, the case of countries such as Italy, or Ireland, with an almost total dependence on imported oil and virtually no indigenous energy sources of any significance to its industrial sector. Rising oil prices make it more difficult, not easier, for alternative energy sources to be deployed. This is because wealth creation, and hence the availability of capital for investment in energy supplies of any kind, depends entirely on the continued availability of oil.

One of the most difficult questions facing the world is how the transition to other sources of energy can take place under conditions of slow economic growth, or recession. One of the paradoxes of the world's energy position at the moment is that the rising price of oil, which signals the need to begin to make a transition away from oil, in fact reduces the total demand for energy and makes the necessity for the transition appear less urgent. There is a real possibility that as oil becomes scarce and dear the industrial world may find itself gently sliding into a decline from which it may not have the resources or the ability to retrieve itself.

Such, then, are some of the dangers which the world is facing as it grapples with varying degrees of concern and urgency with its energy problems. And yet, are things really so bad? Commerce and industry have faltered in the past seven years, but output is still higher than the boom years of the 1960s. The evidence of affluence is everywhere. People may complain that their spending power has declined but they still buy cars, take holidays abroad, install new domestic appliances, and generally feel secure against a return of times as bad as, say, the 1930s when hunger, rather than keeping up the payments on the colour television, was the problem many faced.

Seeing all this, it may be difficult to accept that there are grounds for serious apprehension; or that there is an urgent need to implement measures such as fuel rationing which will drive energy consumption in the industrialized countries down to levels well below today's. The next, and last, chapter will be a brief recapitulation of the argument which has led to this conclusion.

18

Recapitulation

Capture and control of the energy about him has enabled man to emerge from his insignificant position in the biosphere to become the most powerful of its creatures. Coal fuelled the Industrial Revolution; oil carried society beyond it and enabled it to reach the furthest planets; nuclear power has made it possible to make this planet uninhabitable. The balance of military terror is the background against which society must evolve the strategies for its continued survival.

Assuming the absence of a 'discontinuity' caused by large-scale nuclear warfare, the future of society depends finally on the availability of energy. The observations of Frederick Soddy seventy years ago remain valid for today. He said then:

> No one today is ignorant of the part played by energy, not only in science, but in industry, politics and the whole science of human welfare. From the cradle to the grave everyone is dependent on Nature for an absolutely continuous supply of energy in one or other of its numerous forms. When the supplies are ample there is prosperity, expansion and development. When they are not, there is want. Often, it is true, energy appears to play a very subsidiary and indirect part in the development, just as, no doubt, the supply of wind might be looked upon as playing a very secondary role in the music of an organ. The fact remains that, if the supply of energy failed, modern civilization would come to an end as abruptly as does the music of an organ deprived of wind.[16]

The future of society therefore rests on its continuing ability to mobilize the energy resources of the earth. The examination of these, in Part Two, showed that the questions of their magnitude and potential for development are very complex indeed.

The earth is richly endowed with coal – enough for thousands of years' consumption at present rates. Resources of oil and gas are much smaller, each perhaps a twenty-fifth as great. But of much more immediate importance is the fact that these resources are so unevenly distributed. Three-quarters of the world's coal and half the natural gas are in Russia; nearly three-quarters of the world's oil reserves are in the Middle East and North Africa. These resources are outside the control of the main energy-consuming countries of the Western developed world. They are not subject to the 'laws' of supply and demand by which the economies of the major consuming countries are, for the most part, regulated. In comparison with conventional fossil fuels the remaining energy resources of the earth are small, obstinately untappable, or both.

The outlook, therefore, is one of continued dependence on the fossil fuels for a considerable time to come. It would, however, be futile to attempt to predict the exact amount of each of these likely to be produced at any particular date in the future. Any such forecast could only be right by chance. The best that can be done is to try to discover the upper boundaries to possible production and recognize that actual production can be anywhere below this. Taking such an approach, and looking also at the way humanity is expanding, is dependent upon economic growth and even requires, in many cases, an increasing amount of energy merely to stand still, it becomes obvious that with existing levels of waste there is not going to be enough energy to maintain present standards of living in the future. The conclusion is unavoidable; too many arguments point in the same direction.

This approach is essentially much more sceptical than that usually taken by those who produce energy supply and demand forecasts. It does not allow itself to be persuaded into believing that because industrial society would be in trouble without a steady increase in energy supplies such an increase must occur. Neither does it allow itself to be seduced into accepting the argument that, because something is theoretically possible in an ideal world, it will actually happen in the real one. Between the idea and the reality, as T. S. Eliot said, falls the shadow. A rosy technological optimism can picture, and embroider, a future in which nuclear, or even fusion, power stations sprout as freely as mushrooms, solving humanity's energy problems completely and forever. But the beady-eyed sceptic, seeing no sign of any of these things happening, will wait to be convinced, because the consequences of over-estimating energy supplies are much more serious than those of under-estimating them. Norway's problems in deciding what to do with its unexpected riches from the North Sea sit lightly in comparison with those of India, faced with the fact that it can no longer afford to buy all the oil it needs.

Less than a decade ago the well-known geographer Professor Peter Hall could write:

> Population growth has its bad effects as Malthus first reminded us. Looking at India one can see the point; looking at Britain one can see its irrelevance. With every mouth God sends a pair of hands; more relevantly in present conditions, He also sends a brain. An increasing population, in an advanced industrial country, is a younger, better educated, better trained, more adaptable population. It is more capable of taking innovative decisions in production and marketing. The very fact of population growth means an expanding home market and a climate of business optimism. The market will be relatively young, and will create new demands which young entrepreneurs will exploit and sell to the world. If Britain had not possessed a rapidly rising population in the last decade it is less likely we would have developed the Beatles or Mary Quant.[58]

Professor Hall's optimism, more vividly than any pessimism, illustrates the magnitude of the problem of changing attitudes and expectations which confronts the rich countries of the world. Within these there is still a presumption that they are, by divine right, entitled not only to what they already possess but to what they want as well. The assumption is that the pattern not only can, but should, persist: that the young people of Britain, North America and Europe should carry on consuming vast quantities of energy in pursuit of the good life while the young people of India and the rest of the world are content with the Malthusian inevitability of their fate.

In the interconnected world of today such a narrowly limited viewpoint is no longer tenable. Each country will undoubtedly have to accept the final responsibility for its own future. But it cannot ignore the rest of the world while doing so. Most countries now rely so heavily on the international trade in food, energy and other materials that total independence from others is impossible. Communications satellites enable 1 000 million people to see a sporting event such as the World Cup Final; apart from overt military threats, air travel and sophisticated weaponry make every country vulnerable to sabotage and blackmail. It is increasingly difficult for any country to remain complacently unaware of and indifferent to what is happening elsewhere.

If there is not going to be an increase in energy production sufficient to maintain present living standards, let alone increase them, the world is, indeed, facing a difficult time. The indiscriminate growth patterns of the past decades will not be sustained; in many cases they will be reversed. There will be consequent unemployment and economic disruption. Because they no longer control, through the agencies of the international oil companies, the sources of their energy supplies, many of the affluent countries will find their ambitions curbed and their aspirations unfulfilled. In many parts of the developing world the struggle to maintain even the present meagre standards of living will intensify.

It is sometimes fashionable to denigrate the achievements of modern industrial society and compare them unfavourably with the imagined virtues of the simpler life of the past. But literacy, medicine, social services, the liberation of women, the elimination of much of the body-wrecking drudgery of heavy manual work, the growth of an ethical maturity which can find the death penalty abhorrent, are all considerable advances. Of course they have not brought universal happiness. But it is silly to maintain they have not freed many people from a great deal of pain and misery. The most serious objection to them is their exclusiveness. They have been confined to too few people. The challenge to democracy is to retain as much as possible of these good things while managing the great changes required in adjusting to a lower-energy way of life. If it fails, then the collapse into an authoritarianism of either the political right or left seems inevitable.

Making the necessary changes is not going to be easy. Government and corporate planning is usually based on short-term considerations. Politicians want to be re-elected. The oil-company executive is unlikely to find himself promoted if he insists on closing down oilwells to save resources for the future. The modern state has to a considerable extent made itself institutionally incapable of long-term decision-making of a serious kind.

Yet this need not be so. The businessman, trade unionist or politician who accepts the short-term expediencies of public or commercial life, and even considers himself 'realistic' in doing so, will usually adopt an entirely different standard in private life. He will rationalize the most wasteful use of his country's resources by indulging in a vague hope that nuclear power or 'something' will turn up to replace them; at home he will worry about his own old age and his children's future. He will subscribe to a pension scheme and be concerned about the schools in his area. The businessman now investing in an endowment policy for his infant son's university education can be quite complacent about the fact that in twenty years, if present plans are fulfilled, the UK's North Sea oil reserves will be well on the way to depletion. It is now essential the time-span of responsibility which people unashamedly adopt in their personal lives should become the normal standard for public and commercial affairs.

The move to a society which uses less energy, and uses it more efficiently, will require technical and physical changes. Modern science and technology are well equipped to make them. But such changes can only provide part of the answer. The remainder will require a re-evaluation of social goals.

The liberal education systems of the affluent democracies have freed people from the prisons of religious and political dogmatism, but they have not evolved beyond this and become capable of defining goals, except those of never-ending growth in material welfare, for any society. There is an urgent need for profound reflection on education and what it is trying to achieve. If society is going to change, as in most countries it must, then it is worse than useless to train young people in the skills and values of the recent past. In this, as in so many other ways, the affluent countries, however, are fortunate in that they actually have educational systems through which they can work. They also have developed and productive industries which are able to earn them food and energy on the world market. Their populations are close to being stabilized. They have some time in which to deal with their problems.

For the underdeveloped countries the position is much worse. They must obtain more energy very quickly if they are to survive. Their tragedy is that the measures which have, taking the long-term view, fortunately restricted energy consumption in the developed countries bear heavily and destructively on them. The cut-back in world oil production has not made any of it more easily available to those who need it most; it has, instead, placed it further out of their reach. No one can view the problems here without a deep sense of apprehension. Perhaps the greatest hope for the underdeveloped countries is that the affluent countries will find they cannot afford to ignore them. The conduct of the industrial countries in blocking the access of the developing countries to the resources they need whilst at the same time arming them to the teeth is strange indeed.

The outlook is thus undeniably sombre. But the diligent searcher can find some grounds for a cautious optimism. Humanity is not necessarily doomed to crash to destruction in an orgy of overconsumption. In fact, humanity would find it very difficult to do so. Resources, even with the worst will in the world, cannot be exhausted instantaneously. The habit of economy will be forced on people whether they wish it or not, simply because resources will become harder to obtain. If society prepares itself effectively the difficulties this will bring can be overcome. It is possible to cut down on a great deal of the world's present energy consumption and still continue with civilized life. There is enough oil left to provide a frugally minded human race with several more centuries' use of this uniquely useful fuel. But wilful blindness to these facts and an attempt to continue in the old wasteful way, as though energy resources were inexhaustible, will bring frustration and suffering. One thing at least is certain: whatever path is followed, it will be easier if wasteful energy use has been cut to a minimum.

The opportunity to act rationally certainly exists but whether society will do so in practice remains to be seen. This book can offer no set of easy answers or infallible prescriptions. It does no more than begin to provide an outline of the problem, and the choice.

Appendix

Conversion Factors

LENGTH

1 inch = 2·5400 centimetres
1 foot = 0·3048 metres
1 statute mile = 1·6093 kilometres

1 centimetre = 0·3937 inches
1 metre = 3·3281 feet
1 kilometre = 0·6214 miles

AREA

1 acre = 0·407 hectares
1 hectare = 10 000 square metres

1 hectare = 2·4710 acres
1 square kilometre = 100 hectares

1 square mile = 2·5899 square kilometres = 640 acres
1 square kilometre = 0·3861 square miles = 247·1 acres

WEIGHT

1 lb = 0·4536 kilogram

1 kilogram = 2·2046 lb

1 tonne = 1 000 kilograms = 2 204·6 lb
1 Imperial ton = 2 240 lb = 1·0161 tonnes
1 short (U S) ton = 2 000 lb = 0·9072 tonnes

VOLUME

1 cubic foot = 0·0283 cubic metres
1 Imperial pint = 0·5682 litres

1 cubic metre = 35·3147 cubic feet
1 litre = 1·7600 Imperial pints

1 Imperial gallon = 1·2009 U S gallons = 4·5461 litres
1 U S gallon = 0·8327 Imperial gallons = 3·7854 litres
1 barrel = 42 U S gallons = 34·97 Imperial gallons = 159·0 litres

ENERGY

1 kilowatt-hour	= 1·34 horsepower-hours
	= 1·98 × 10^6 foot pounds
	= 3 412 British thermal units
	= 859·845 kilocalories
	= 3 600 kilojoules
	= 3·6 megajoules = 0·0036 gigajoules
	= 0·034 therms
1 gigawatt-hour	= 1·0 × 10^3 megawatt hours
	= 1·0 × 10^6 kilowatt hours
	= 1·0 × 10^9 watt-hours
1 gigajoule	= 277·78 kilowatt-hours

POWER

1 kilowatt = 1·34 horsepower

TEMPERATURE

1 Kelvin = 1 scale degree Celsius = 1·8 scale degrees Fahrenheit

For conversion:

$$°C = (°F - 32) \times \frac{5}{9}$$

$$°F = \left(°C \times \frac{9}{5}\right) + 32$$

0 K ('absolute zero') = −273·15°C

OIL INDUSTRY UNITS

1 tonne crude oil	= 7·3 barrels (average)
	= 256 Imperial gallons
	= 301 U S gallons
1 barrel per day	= 50 tonnes per year (average)
1 tonne motor spirit	= 8·45 barrels (average)
1 tonne fuel oil	= 6·70 barrels (average)

UNITED NATIONS STATISTICS USE THE FOLLOWING CONVERSION FACTORS:

1 tonne of coal is equivalent to:

1·0 tonne of brown coal in New Zealand
2·0 tonnes brown coal in Czechoslovakia, France, People's Republic of Korea, Albania, Austria, Bulgaria, Hungary, Italy, Portugal, Spain, USSR, Yugoslavia
3·0 tonnes of brown coal in other countries
2·0 tonnes of peat
0·77 tonnes of crude oil
0·66 tonnes of petroleum products
750 cubic metres of natural gas
8 000 kilowatt-hours

The output of hydro and nuclear power stations in UN figures for primary-energy consumption is counted in terms of its calorific value, i.e., 8 000 kilowatt-hours = 1 tonne of coal equivalent.

STATISTICS COMPILED BY BRITISH PETROLEUM USE THE FOLLOWING:

1 tonne of oil is equivalent to:

1·5 tonnes of coal
4·9 tonnes of lignite
3·3 tonnes of peat
1 167 cubic metres of natural gas
415 therms
41×10^6 British thermal units
12 000 kilowatt-hours

The output of hydro and nuclear power stations is measured in terms of the notional amount of fossil fuel required to produce this electricity at a generating efficiency of 30 per cent, i.e., an electricity output of 12 000 kilowatt-hours counts as a primary energy input of 3·0 tonnes of oil.

1 cubic metre of natural gas is equivalent to:

10·47 kilowatt-hours.

Lists of Figures and Tables

Figures

Tables

List of Sources Quoted

Sources of tables and figures are noted where they appear. The following sources are listed in the order in which they first occur.

1. Thomas Robert Malthus, *An Essay on the Principle of Population*, first edition, 1798; Penguin, 1970.

2. H. G. Wells, *The World Set Free*, first edition, 1914; Collins, 1956.

3. K. Marx, *Capital* quoted in T. B. Bottomore and Maximilien Rubel, eds., *Selected Writings in Sociology and Social Philosophy*, Penguin, 1967.

4. J. Stanley Clark, *The Oil Century: From the Drake Well to the Conservation Era*, University of Oklahoma Press, 1968.

5. Herbert S. Jevons, *The British Coal Trade*, first edition, 1915; David & Charles, 1972.

6. Frederick Alderson, *Bicycling: A History*, David & Charles, 1972.

7. 'Exploring Energy Choices' (a preliminary report of the Ford Foundations's Energy Policy Project), Washington, 1974.

8. Daniel Bell, 'Notes on the Post-Industrial Society', quoted in Nigel Cross, David Elliot and Robin Roy, eds., *Man-Made Futures: Readings in Society, Technology and Design*, Hutchinson Educational in conjunction with the Open University, 1974.

9. E. L. Trist, 'The Structural Presence of the Post-Industrial Society', quoted in *Man-Made Futures*... (see Source 8).

10. J. D. Bernal, *Science in History*, Volume 2, Penguin, 1969.

11. Frederick Soddy, *Cartesian Economics*, Henderson, 1922.

12. Frederick Soddy, *Wealth, Virtual Wealth and Debt*, Allen & Unwin, 1926.

13. Howard T. Odum, *Environment, Power and Society*, Wiley-Interscience, New York, 1971.

14. Malcolm Slesser, 'The Energy Ration', *Ecologist*, May 1974.

15. J. Darmstadter, J. Dunkerly and S. Alterman, *How Industrial Societies Use Energy*, Johns Hopkins University Press, 1977.

16. Frederick Soddy, *Matter and Energy*, Williams & Norgate, London, 1912.

17. M. K. Hubbert, chapter on energy resources in Committee on Resources and Man, National Academy of Sciences National Research Council, *Resources and Man*, W. H. Freeman, San Francisco, 1969.

18. World Coal Study, *Coal – Bridge to the Future*, Ballinger, Cambridge, Mass., 1980.

19. E. N. Tiratsoo, *Oilfields of the World*, Scientific Press, Beaconsfield, Bucks., 1973.

20. 'A Blueprint for Survival', *Ecologist*, January 1972.

21. P. R. Odell, 'The Future of Oil: A Rejoinder', *Geographical Journal*, June 1973.

22. Wallace F. Lovejoy and Paul T. Homan, 'Methods of Estimating Reserves of Crude Oil, Natural Gas and Natural Gas Liquids', *Resources for the Future*, Johns Hopkins Press, Baltimore, 1965.

23. Energy Research Unit, Queen Mary College, London, 'World Energy Modelling: The development of Western European oil prices', *Energy Policy*, June 1973.

24. S. C. Ells, *Recollections of the Athabasca Tar Sands*, Department of Mines and Technical Surveys, Ottawa, 1962.

25. *Energy Conservation: Ways and Means*, Future Shape of Technology Foundation, The Hague, 1974.

26. *An Energy Policy for Canada*, Volumes I and II, Ministry of Energy, Mines and Resources, Ottawa, 1973.

27. *Survey of Energy Resources 1974*, US National Committee of the World Energy Conference, New York, 1974.

28. Lawrence M. Lidsky, 'The Quest for Fusion Power', *MIT Technology Review*, January 1972.

29. J. C. Emmett, J. Nucknolls and L. Ward, 'Fusion Power by Laser Implosion', *Scientific American*, June 1974.

30. G. L. Kulchinski, 'Fusion Power – An assessment of its impact on the USA', *Energy Policy*, June 1974.

31. H. C. Hottel and J. B. Howard, *New Energy Technology: Some facts and assessments*, Massachusetts Institute of Technology Press, 1971.

32. *Energy Conservation: A study by the Central Policy Review Staff*, Her Majesty's Stationery Office, 1974.

33. S. H. Salter, 'Wave Power', *Nature*, 21 June 1974.

34. Thomas B. Johansson and Peter Steen, *Solar Sweden*, Secretariat for Future Studies, Stockholm, 1978.

35. D. E. Earl, *Forest Energy and Economic Development*, Clarendon Press, Oxford, 1975.

36. Donald P. Snowden, 'Superconductors for Power Transmission', *Scientific American*, April 1972.

37. R. F. Post and S. F. Post, 'Flywheels', *Scientific American*, December 1972.

38. J. K. Dawson, 'The Prospects for Hydrogen as a Fuel in the UK', *Atom*, May 1974.

39. L. B. Escritt, *Sewage Treatment: Design and Specification*, Contractors Record, London, 1950.

40. L. John Fry, *Practical Building of Methane Power Plants for Rural Energy Independence*, D. A. Knox, Andover, Hants, 1974.

41. Ariane van Buren (Ed.), *A Chinese Biogas Manual*, ITDG Publications, London, 1979.

42. D. H. Meadows *et al.*, *The Limits to Growth*, Universe, New York, 1972.

43. Chauncey Starr, 'Energy and Power', *Scientific American*, September 1971.

44. *Energy for the Future*, Institute of Fuel, London, 1973.

45. Conservation Commission of the World Energy Conference, Executive Summaries *World Energy Resources 1985–2000*, IPC Science and Technology Press, 1978.

46. Centre of Housing Building and Planning, UN Department of Economic and Social Affairs, *Human Settlements: The Environmental Challenge*, Macmillan, 1974.

47. Colin Clarke, *Population Growth and Land Use*, Macmillan, 1967.

48. H. Ozbekhan, *The Triumph of Technology*, quoted in *Man-Made Futures*... (see Source 8).

49. Gerald Leach, 'The Impact of the Motor Car on Oil Reserves', *Energy Policy*, December 1973.

50. International Energy Agency, *Energy Policies and Programmes of IEA Countries: 1978 Review*, OECD, Paris 1979.

51. G. Leach, C. Lewis, F. Romig, A. van Buren and G. Foley, *A Low Energy Strategy for the United Kingdom*, IIED, 1979.

52. Lewis Chester, *Sunday Times*, 6 January 1972.

53. Rachel Carson, *Silent Spring*, Houghton-Mifflin, New York, 1962; Penguin, 1965.

54. David Dickson, *Alternative Technology*, Fontana/Collins, 1974.
55. Godfrey Boyle, *Living on the Sun*, Calder & Boyars, 1975.
56. Aldous Huxley, foreword to *Brave New World*, Triad, 1946.
57. Amory B. Lovins, *Soft Energy Paths*, Penguin, 1977.
58. Malcolm Chisholm, (ed.), *Resources for Britain's Future*, Penguin, 1972.

Index

More about Penguins and Pelicans

For further information about books available from Penguins please write to Dept EP, Penguin Books Ltd, Harmondsworth, Middlesex UB7 0DA.

In the U.S.A.: For a complete list of books available from Penguins in the United States write to Dept CS, Penguin Books, 625 Madison Avenue, New York, New York 10022.

In Canada: For a complete list of books available from Penguins in Canada write to Penguin Books Canada Ltd, 2801 John Street, Markham, Ontario L3R 1B4.

In Australia: For a complete list of books available from Penguins in Australia write to the Marketing Department, Penguin Books Australia Ltd, P.O. Box 257, Ringwood, Victoria 3134.

INTERNATIONAL PEACEKEEPING

United Nations forces in a troubled world

Anthony Verrier

Irish squaddies are shot in a squalid encounter in the Middle East and the newspapers are pious for a week until the next bit of international 'news' displaces them. Unlovely acronyms, ONUC, UNFICYP and UNIFIL are trotted out for a bemused and bored public, but they conceal a real and urgent, a terrifying and crucial, a randomly violent half-war.

These are the visible signs of an astonishing and largely hidden military crusade. It is the struggle for peace waged, sometimes uncertainly but unceasingly, by the United Nations Peacekeeping Forces. The strange armies made up of amazing mixtures of people speaking a multitude of different languages under, for most of them, an alien command, keep warring factions apart.

This book describes the ways these unsung heroes work, shot at by all but unable, for the most part, to shoot back. Verrier recounts their history, their defeats and their victories. He shows how the strategies work or are improvised and, above all, how vital is their role in international politics.

A Pelican Original

DISASTERS

The anatomy of environmental hazards

John Whittow

A disaster is categorized as such only when there is loss of human life or property. A volcanic eruption, a massive earthquake or a landslide in barren terrain is not a disaster. Yet over the last four or five years rarely a month has gone by without a major disaster being reported from somewhere in the world – the Bangladesh cyclone; the Huascaran avalanche in Peru; the Nicaraguan earthquake; the Turkish earthquake; the Honduras floods spawned by Hurricane Fifi; the Sahel drought of North Africa. Hundreds of thousands of lives have been lost. The disastrous year of 1976 appeared to be the final cataclysm, when earthquake shocks, volcanic eruptions, tidal waves, hurricanes, floods, blizzards and drought combined to give the impression that the earth was in turmoil, that the day of reckoning was at hand.

What is going on? Is it possible to predict such events or to alleviate their effects? John Whittow not only summarizes the disasters which have caused such widespread death and destruction but also explains, as far as possible, why they occur and whether man is partly to blame.

A Pelican Book

INFLATION

A guide to the crisis in economics

SECOND EDITION

J. A. Trevithick

Inflation, accompanied by economic disarray and popular despair, continues to offer a major challenge to the credibility of economic science.

As the flow of explanations bids fair to match the growth in the money supply, Dr Trevithick gives – in this Pelican Original – an introductory guide to the problems, the jargon and the panaceas.

The characters in his drama include Milton Friedman, Keynes, Hayek and the 'New School' of Cambridge, while the plot encompasses the Phillips Curve, trade-union power, helicopter money, rational expectations, floating currencies, incomes policy, import controls and indexation.

In this new edition the author brings us up-to-date with such things as Friedman's Nobel Lecture in 1977, and Lord Kaldor's injection of a new convincingness to the cost–push theory, as well as analysing the latest manifestations of current monetarist orthodoxy in both Britain and North America. He also tells us his view of what will have to be done.

A Pelican Book

PROGRESS FOR A SMALL PLANET

Barbara Ward

Pollution, diminishing oil reserves and the chasm between the rich and poor nations dominate our headlines. Some people claim such crises are insoluble, others that the new technologies hold the answers. *Progress for a Small Planet* accepts neither view, but shows how new attitudes backed by new, more conserving, technologies can take us beyond these crises. Barbara Ward outlines the planetary bargain between the world's nations that would guarantee every citizen the right to freedom from poverty and keep our shared biosphere in good working order.

A Pelican Book